Mineral Prospecting Manual

Mineral Prospecting Manual

Coordinated and produced for the Service Formation-Coopération du BRGM (Bureau de Recherches Géologiques et Minières) by:
Jean-Bernard Chaussier

Consultant: **Jean Morer**

Translated by
Service Traduction du BRGM

Elsevier
New York · Amsterdam · Oxford

Revised and updated 1987

English language edition first published 1987 by North Oxford Academic Publishers Ltd, a subsidiary of Kogan Page Ltd, 120 Pentonville Road, London N1 9JN in association with Bureau de Recherches Géologiques et Minières, BP 6009, 45060 Orléans cedex, France

Sole distributors in the United States and Canada:
ELSEVIER SCIENCE PUBLISHING Co., INC.
52 Vanderbilt Avenue, New York, New York 10017

Library of Congress Cataloging-in-Publication Data

Chaussier, Jean-Bernard.
Mineral prospecting manual.

Translation of: Manuel du prospecteur minier.
Bibliography: p.
1. Prospecting—Handbooks, manuals, etc. I. Morer, Jean. II. Title.
TN270.C41813 1987 622'.1 86-4602

ISBN 0-444-01079-9

Printed and bound in Great Britain

Contents

Foreword

The importance of mineral raw materials and of ores particularly, both for consumer and producer countries, cannot be over-emphasized. It has become vital to discover new metal deposits and the search for them has involved the use of increasingly sophisticated techniques such as geochemistry and remote sensing. However, while these techniques are important and their development is vital, the field research stage still remains crucial and the success of such research depends on the field prospector to a large extent.

It is therefore important that a single manual of all the rules, methods and 'tricks of the trade' used in mineral prospecting is available for future generations of prospectors, some of them complete beginners, who are still going out into the field to explore for minerals. One of the authors of this book, J Morer, had already proposed such a manual, in 1959, with J Sagatsky. Only a few copies were published, but the work was a great success. Since then, old methods have been improved and others have been introduced, so an up-to-date manual becomes an absolute necessity.

This manual is a collective work. J-B Chaussier drew on his many experiences of the hazards of being a mining geologist and picked and coordinated the areas of interest to be included in the book, including topics that may seem obvious to some readers. He was helped by J Morer who, at the end of a career which began at the BRGM just after the Second World War, was happy to spend long hours advising us and to make the benefit of his own experiences available to our readers.

The introduction written by J P Dumas explains how the techniques described correspond to the various phases of mining research. J-B Chaussier describes the physical preparation for prospecting (Chapter 1); he also wrote Chapter 3 on the prospecting of alluvial and eluvial deposits. The material on diamond prospecting was provided by J-C Michel (Chapter 3). J Marchesseau gives a detailed study of prospecting coastal formations (beach sands, dunes, etc.). The location, examination and sampling of outcrops are described by J Letalenet in Chapter 2, which is devoted to the subject now generally called hammer prospecting; this type of prospecting remains the basis of any mineral exploration by geologist or prospector. The use of more sophisticated research methods for non-outcropping occurrences is described by L Viallefond for geochemical prospecting (Chapter 4) and by R Millon for geophysical prospecting (Chapter 5). These methods are becoming increasingly more important as the number of undiscovered outcropping ores falls and exploration becomes more oriented towards hidden deposits. In Chapter 6 on drilling in mineral exploration, B Villeneuve wrote the section on drilling techniques and equipment, while sampling was described by J Deroubaix for core drilling and by J Testard and J Morer with the help of R Trinquard for percussion drilling. P Battut drew on his own experience in excavation to describe these techniques for prospectors who need to use them. The description of different sampling procedures in prospecting work is the responsibility of J-P Prouhet, J-B Chaussier and J Morer (Chapter 7).

I should like to thank and indeed congratulate all the members of the Publication Department at the BRGM, as well as the Cooperation Department including its head, Jean-Claude Napias, who collated the material into a manual for researchers wishing to familiarize themselves with French techniques.

Jean Ricour
Director of the BRGM
November 1986

Introduction

General mineral prospecting is essentially a natural experimental process. It is carried out in phases that are distinguished by the areas involved and the techniques used, and consequently also by the different human, material and financial means required.

In a good mineral prospecting programme there should always be a reasonable relation between the risk being taken and the estimated profit. To achieve this is not easy and it is clear that a critical examination of technical, economic and financial data is imperative at the end of each phase. The results obtained must be carefully considered in order to appreciate the risk involved and to define future aims. A decision on whether or not to continue prospecting will follow and, if the decision to go ahead is taken, new objectives and the means for achieving them have to be defined. As the work advances, so these decisions become of greater consequence and the danger of passing on to the next phase without having fully answered the questions raised by the last one cannot be over stressed. W G Peters was right in saying that 'the greatest financial losses occur most often, not in exploration but in exploitation when this turns out to be unprofitable after investment decisions have been made on the basis of an incomplete geological study. The decision to consider an anomaly as a target area is less critical than that of considering a badly studied area as a deposit.'

At each stage of mineral prospecting the decision to continue or stop depends, above all else, on qualitative and quantitative geological and mineralogical data – dimensions, grades – gathered on the show or deposit being studied. Programmes of investigation anticipate the use of particular methods to determine the points of intersection, to check them, reconnoitre the mineral deposits discovered and to evaluate any ore bodies. It is of course very important to choose the appropriate methods but it is equally essential to be certain of the validity of the results obtained because their application influences the decisions made.

Each prospecting method must be implemented with extreme care, from the simple examination of outcrops or alluvial panning to the gathering of cuttings from percussion holes or to the collection of samples from excavations. It is usually the prospector who is responsible for such work, and it is to the prospector that this manual is directed. It is intended simply to provide as complete as possible a series of guidelines for the work he will have to do and the precautions he will have to take. Professional prospectors have contributed to this book, and have indeed written parts or all of some of the chapters.

The prospector cannot be a universal expert who understands everything about the use of the most complicated and sensitive geophysical prospecting equipment, but he must be able to apply simple geophysical techniques to ensure that the information collected in the field is complete, and so avoid the additional costs of sending out a specialized team.

Similarly, the prospector can also be trained to use light- weight drills or to excavate short reconnaissance drifts or shallow exploration shafts. He will find the information he needs, in particular about blasting, timbering and ventilation in this manual. His most important role, however, will be not to take the place of the specialists, namely the drillers and miners, but to direct and supervise their work and to take the samples necessary to obtain good geological and prospecting information about the deposit. Sample collecting is without a doubt the most important and also the most delicate operation at every stage of mineral prospecting, particularly in the final phase, and for this reason the prospector should take great care when collecting samples to be sent for analysis, and when quartering to reduce the original amount. These operations are also described in different chapters in the manual.

As well as geologists and prospectors, practising geophysicists, drilling engineers and mining engineers have also contributed to this book. This manual is thus the work of a large broad-based group reflecting the BRGM's experience in mineral prospecting in all latitudes and all climates.

Chapter 1

Preparation for and organization of a mineral prospecting expedition to a tropical area

This chapter constitutes a series of guidelines for operations to be effected and the materials necessary for carrying out a geological field survey in a desert, tropical or equatorial climate. Lists of materials used by the BRGM are given in Appendix 1. They are as complete as possible, but may have to be modified to take into account particular circumstances, as well as regional, climatic and financial conditions.

1 Preparations

1.1 Technical information

1.1.1 REMINDER

A newly appointed employee at the BRGM, or in any company, should first of all be told about the organization, its history, its aims and above all its resources. A detailed visit to all its departments and discussions with the heads of the various supporting services such as physicochemical analysis laboratories, petrography, geochemistry and graphic arts are advised. Some time spent, for example, in the sample reception shed will enable the prospector to see what the best form of packaging is so as to avoid accidents, confusion, losses and any mishap which could affect preparations for analysis and results.

1.1.2 OBJECTIVES OF THE FIELD STUDY

The head of the expedition should be issued with instructions accompanied by a technical data sheet (see Appendix A1) concerning the programme. This defines the objectives and means of the expedition and also summarizes the information available. The head of the expedition must obtain information from all the specialists concerned about the particular problems facing each member of the group; he should then allocate the various preparatory tasks to them.

If it is the first time the company has worked in a particular country, full information on geographical, climatic, political, ethnological, legal and human conditions in the country must be obtained and the whole group should be made familiar with it, so that breakdowns in relations with the authorities of the country can be avoided, at both the social and the political levels.

1.1.3 THE STATE OF THE ART

A general study of the work carried out in the country and of previous geological reports should be carried out; where appropriate contact should be made with earlier heads of expeditions so that tasks can be taken over from previous expeditions, or with geologists who have already worked in the country and may be able to help with background research. After careful selection copies of certain essential documents should be made and taken into the field.

The Documentation Department at the BRGM in Orleans collects all the literature published in the world on Earth Sciences. In collaboration with its French and European partners, the BRGM has thus established a bibliography database containing over 450,000 references. The database covers worldwide geological information published since 1968 and French geology over the last two centuries. It can be consulted for the production of monthly bibliographies or to obtain reference lists covering several years, while books and periodicals can be consulted in the library.

1.1.4 MAPS

Maps of the work area should be assembled, in sufficient numbers for the whole group and to last throughout the expedition. Enlargements might perhaps be necessary for certain areas and more detailed information added to them, drawn from older maps or from references in the bibliography. The location of bases for geochemical work should also be planned.

1.1.5 PHOTOGEOLOGY

Aerial photographs of the work area should be assembled; they should be closely studied and photogeological maps prepared in sufficient numbers including reserves as necessary; stereoscopes need to be provided. Specialists should be consulted concerning remote sensing work on the most important zones or those which need to be examined and/or identified first. The map, combined if possible with existing maps enlarged to a scale of 1:50,000 or 1:25,000, can be used as a route-planning map during strategic prospecting, or, combined with a cadastral map, can be used for tactical reconnaissance.

1.2 Administration

1.2.1 GENERAL CONTEXT

Another aspect which is too often neglected during expedition preparation is administration. The approach adopted varies from one company to another. It is essential to find out about administrative structure and procedure and to spend some time with a member of the administrative staff who will provide the documents necessary for the expedition and explain their often pedantic style.

1.2.2 REGIONAL CONTEXT

Local administrative procedures also must be studied carefully, particularly from the point of view of the labour legislation in force in the country where the expedition is to take place. These questions can generally be best sorted out on the spot by the local representative of the company, if there is one; if not, a member of the group should contact the authorities in the region to find out about:

1) The hiring of salaried personnel (drivers, labourers, mechanics, storekeepers, etc.), the standard layout of wageslips, wage rates, salary structures, tax deductions, pensions, etc., what allowances are payable and what the beneficiary has to produce to prove entitlement to them (eg forms used by government departments and local banking procedures).
2) Matters relating to medical examinations: when are they necessary, who carries them out, who pays for them.
3) Labour legislation: hours, bonus schemes, conditions of dismissal, relevant regulations.
4) The hiring of employees paid by the hour (site labourers, porters, camel drivers, boatmen, etc.), hourly rates and procedures for payment (daily, weekly, monthly or according to any other scheme agreed with the local labour officials that is compatible with the activities of the expedition).

1.2.3 ACCOUNTS

Agreement must be reached on accounting procedures, availability of funds and regular remittances from the home country either with the local representative or, if there is none, with a bank which is prepared to provide money on a fixed date. The financial responsibility of the expeditions head must be defined before departure. Bearing in mind the difficulty of getting change in some isolated villages, remember to take notes in small denominations and to have plenty of small change.

1.3 The journey

Before leaving the home country it is essential to find out exactly what will be available in the country of study, so as to be able to take anything that is not available locally, and in general get properly organized before departure. When ordering equipment, think about delivery times, which can sometimes be very long, customs clearance procedures and the visas necessary. There may be problems taking hunting weapons or ammunition into certain countries.

Think about vaccinations and read about the diseases you could come into contact with. Institutions specializing in tropical medicine are the best sources of information. Take whatever medicines you might need, see your doctor and have a dental check-up.

How can you get to where you are going; how much baggage are you allowed? Find out about arrival times, local currency, the possibility of using foreign currency (dollars, sterling, francs, etc.) abroad? Will you be met and will there be sufficient accommodation from the first night? Who should you contact amongst the local authorities? Which local authorities are the most important? Contact the embassy. Where will you find a bank? Are you authorized to make bank transfers and withdrawals? Also think about registering the members of the expedition at your consulate to enable local contact. It is always advisable to send someone ahead to look around, and to leave someone behind to check that all the equipment is returned.

2 The expedition itself

2.1 The camp

2.1.1 CHOICE OF BASE CAMP

Find the most favourable area for setting up the base camp. If possible, it should be near a communication route, a supply centre and a water hole. Aim to set up the camp in buildings where equipment can be locked in a store.

Try to choose a pleasant, clear site which is shaded and can be enclosed if necessary. Find out about regional medical and hospital facilities so that in the event of an accident or illness no time is lost. If there are no nearby facilities, locate landing ground nearby and runway beacons, and familiarize yourself with the means of raising the alarm, runway markings and ground-to-air signalling codes, etc.

In desert areas, if there are no settlements nearby, it is advisable to use caravans or other mobile accommodation with lights, air conditioning and a generator. This makes it easier to sleep at night with the result that the group works more efficiently during the day.

2.1.2 FITTING OUT THE BASE CAMP

The base camp must be as comfortable as possible and equipped with a gas or kerosene (paraffin) refrigerator and freezer, gas cooker, beds, bedding, sheets, pillows, mosquito nets, armchairs, good gas, kerosene or electric lighting and a toilet area with good showers and isolated hygienic toilets which are regularly disinfected and moved after being filled in.

Large drawing tables that can be set up on the spot from planks or plywood boards should be available; equipment stores with shelves and a well-protected store for provisions, resistant to ants and rats, etc. by sump-oil or diesel fuel impregnated into the wood and applied every day, or by an insecticide should also be available.

Fuel supplies should be well-ventilated and protected from direct sunlight by a tarpaulin, neither too close to the camp, nor so far away as to encourage thieves. There should be sloping loading facilities for fuel barrels and a pit that enables the underside of vehicles to be maintained.

If there are no permanent buildings, use large tents with a double roof and large canopy. Dig a channel around the tents and fill it with diesel fuel.

2.1.3 FLY-CAMPING

Equipment is lighter and reduced to a minimum; the more limited the transport, the more limited the equipment will be. Tents are less spacious and beds smaller, sometimes just a sleeping bag. Two canvas covers, a mattress or hammock and mosquito net, one saucepan and some candles, plus a change of clothes, would, if pushed, be all the camping equipment necessary for one person. It is a good idea to use the 'umbrella' or 'igloo' type tent with a diameter of about 2m on the ground and a height of 1.20m, which can be put up very quickly, without pegs or stays, on any ground, rocky or muddy, on matting or on a platform in a tree. These tents are designed to be completely watertight, so that it is always dry inside and a mosquito net is not necessary, and they can be heated by a single candle. There are also ultra-light bivouac tents on the market, weighing 1kg and measuring 220 × 140 × 100cm, but made of nylon.

Before setting up camp, think about possible pollution. To avoid daily loading and unloading of equipment and tents, it is preferable to use a camping bus for a travelling camp in a Sahel or Sahara desert area. The camping bus can also be an excellent base for a fixed site (for drilling, tactical prospecting, etc.).

2.2 *Means of transport*

What means of transport are available for the expedition? What condition are they in? Whether for river, land or air travel, are they suitable for the conditions of work? Will it be possible to find spare parts locally? The make of vehicle must be chosen so that proper after-sales service and any assistance necessary on the expedition can be relied upon and a stock of replacement parts guaranteed.

2.2.1 MOTOR VEHICLES

A maintenance handbook and a list of essential spare parts for a vehicle may be more difficult to obtain abroad than at home. A minimum stock of spare parts should be obtained and sent out, but customs regulations, formalities, import taxes and time necessary for delivery must be taken into account. These vehicles are generally standard models which are not always suitable for the climate in the area of study. It is therefore advisable to anticipate any modifications that are necessary to take account of the climate and the uses to which the vehicle will be put (sce Appendix A2 for safety instructions for vehicles travelling in a desert area).

The following is a useful checklist:

1) spare petrol tanks which should be flat and compartmentalized for long journeys in desert areas (two 80 litre tanks to guarantee a 1000km range);
2) water tank or jerrycan (placed at the back of the vehicle) for long journeys in desert areas (never put a petrol jerrycan in the front of a vehicle);
3) door pockets for documents;
4) sun visor;
5) spotlight on the driver's side for night driving over rough terrain (not strictly allowed in desert areas);
6) a driver's compass placed above the steering wheel;
7) a compass for the navigator on the passenger's side with a map-board for use in desert areas;
8) engine thermometer and oil pressure indicator are essential in hot areas;
9) good lighting in the cabin (useful when the vehicle is stationary);

10) manual throttle control (useful in desert areas for automatic driving);
11) exterior bottle carriers to keep drinks cool;
12) canvas covers made from old tents, fixed along the outside of the car, and forming an awning for the campsite at night;
13) roofrack, on the front of the roof to balance the load on the axles;
14) spare spring on the front bumper;
15) spare tyre on the bonnet, where it is easier to reach though visibility is impaired, shovels and axes on the front wings;
16) a double cabin roof is a protection against heat;
17) metal sheets fixed on the sides give protection in sandy or muddy areas;
18) two batteries should be carried, especially if the vehicle is equipped with a radio and an alternator for charging;
19) suspension should be reinforced and steering should be power assisted for rough terrain;
20) towline.

A winch on a vehicle can be very useful in rough terrain. Certain spare parts can easily be sorted out before departure. If there is no mounted winch, a hand-operated winching device can be of great help when a vehicle is stuck in mud or sand. Depending on where the vehicle was built, check that tools have been designed on the same measurement system as the parts (metric bolts are hard to remove using imperial tools). It may be a good idea to set up two coil circuits side by side. Remember that hoses and rubber joints crack very quickly in heat, and glue should be carried in case of breaks.

Mopeds can be used in Europe, in desert areas, in light forests and on plains and enable prospecting to be quickly done from a base camp, using tracks inaccessible to vehicles. This allows prospectors to collect more samples in less time. They can also be used for carrying mail over short distances. However, be sure to take enough fuel for the journey, a spark plug socket and a puncture repair kit. (See Appendix 3 for the use of vehicles in desert areas.)

2.2.2 RIVER CRAFT

Check the boat's characteristics (capacity, buoyancy, etc.) and whether it is suitable for the river in question, from the point of view of depth of vegetation, presence of shallow water or rapids, the need for portage, etc. Remember to take enough fuel for the size of the engine and essential spare parts, propellor joining pins, pin driver, spark plugs and spark plug sockets, two full tanks of fuel, some made-up mixture in a jerrycan, two paddles, a spare starting rope, a mooring rope, some rags, a syringe and oil for the sump, propellor and grease box (see Appendix 1).

It is worth bearing in mind the range of an outboard motor with a 23 litre tank (standard model):

1) 40hp motor – at maximum speed: 1h 15, at minimum speed: 2h 15;
2) 28/25hp motor – at maximum speed: 1h 45, at minimum speed: 3h;
3) 18hp motor – at maximum speed: 2h 30, at minimum speed: 3h;
4) 10hp motor – at maximum speed: 4h, at minimum speed: 6h;
5) 5hp motor – at maximum speed: 6h, at minimum speed: 11h.

A dugout canoe is often a better choice than an unsuitable metal boat.

2.2.3 ANIMALS

As well as mechanized vehicles, a prospector may find it easier to use animals as a means of transport: horses, donkeys, mules or oxen – pack, ridden or harnessed – camels, lamas or dogs. Saddles, packsaddles, or harnesses vary according to the country, but it is always advisable to use local equipment. An animal accustomed to a certain type of saddle or harness will find it difficult to adapt to another, and the advice of a guide or the animals' owner should be followed. If no veterinary assistance is available, make sure that the animal has no suppurating wounds, particularly in the places where packsaddles, saddles, harness straps or harnesses are fixed.

The guide or owner will take care of the animals, but remember that it is advisable to take only the absolute minimum when travelling by camel or horse, ie a folding mattress, a duvet, a blanket and a groundsheet, which can also be used to make the saddle comfortable, as bedding; a kettle for tea, and a billy can for all other cooking. Do not overload yourself with tinned foods but eat as much local produce as possible. Before leaving check the condition of the saddles, girths and harnesses and take whatever is necessary to maintain and repair them (eg needle and thread, straps, possibly an eyelet riveting machine). Remember to grease or waterproof them according to the weather in the area: hot or cold, dry or wet.

2.2.4 AIRCRAFT OR HELICOPTERS

Leave aircraft maintenance to the pilot. In a helicopter operation make arrangements for picking up passengers, clearly identified by bright orange overalls for location on ground and if possible equipped with a walky-talky radio even of the very short range childs toy type, a maximum range of 100m is enough. If the helicopter cannot land to drop the group and each member has to jump, make sure that the ground is firm by first throwing a package on a rope before jumping yourself. Look out for quicksand and swamps or lakes covered with vegetation.

The use of a helicopter makes it necessary to clear a landing area 50m by 100m, in the direction of the prevailing wind, as shown in the **Fig. 1.1**. If such a place already exists and is clear of undergrowth, you should nevertheless indicate the landing site using a 3m by 3m triangle and the direction of the wind by a T-shaped sign on the ground. If possible the wind direction should be indicated by a permanent wind sock, as there may not always be anyone on hand to guide aircraft in, set out the T and light a smoke-bomb. If the ground is marshy, a mat can be made of wood and branches.

When signalling with a smoke-bomb, it must be placed at the downwind end of the landing area. If there are no cartridges, a smoke signal can be made with a cloth soaked in engine oil or from a food tin filled with sand, petrol and motor oil or from a piece of tyre soaked in petrol. If you have no matches or lighter, you can use a rag soaked in petrol and a screwdriver to short circuit a battery to make a fire. A geologist's lens or possibly a pair of glasses or watch face can be used to ignite leaves and twigs. For cigarette smokers the 'solar lighter' (Kreops lighter) can always be used in daytime in desert areas.

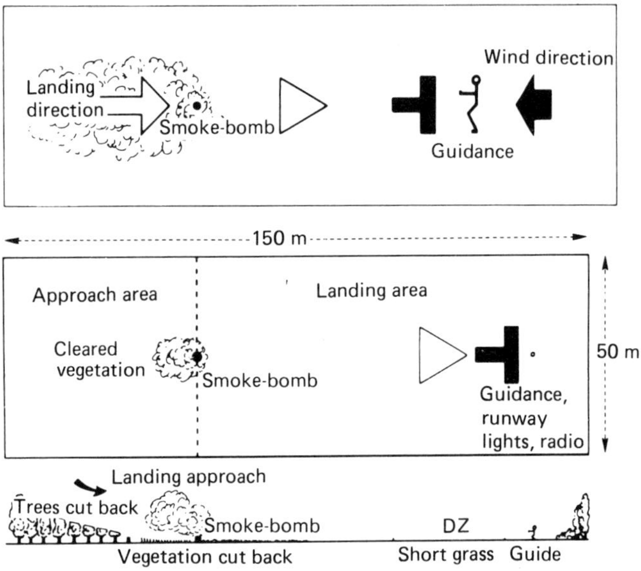

Fig. 1.1. *Landing runway for helicopter.*

2.3 Communications

Communication is of fundamental importance for any expedition, and attention must be paid to it both before departure and on arrival to obtain permission from the local authorities. There are three types of communication to be considered:

1) direct contact, by car, chartered aircraft, boat, canoe or runner for mail and supplies;
2) official post office mail, telephone or telegram, using the normal system, which depends on the location of post offices and their opening hours and on the official air mail services;
3) private radio, which requires special permission, a schedule and a procedure to be followed. Ask for information at the head office of the telecommunications department in the country concerned.

The types of communication must be specified before departure and strictly adhered to.

2.3.1 RADIO PROCEDURES

Generally, to be able to send radio messages the operator must hold a national radio operator's licence, which requires certain knowledge, particularly concerning operating procedures and repairs to equipment. Certain countries tolerate non-qualified operators so long as they follow international procedure. Make enquiries beforehand and learn how to use the equipment correctly.

Voice radio letter code			Morse code		
A Alpha	J Juliet	S Sierra	A .–	M – –	Y – . – –
B Bravo	K Kilo	T Tango	B – . . .	N – .	Z – – . .
C Charlie	L Lima	U Uniform	C – . – .	O – – –	1 . – – – –
D Delta	M Mike	V Victor	D – . .	P . – – .	2 . . – – –
E Echo	N November	W Whisky	E .	Q – – . –	3 . . . – –
F Foxtrot	O Oscar	X X-ray	F . . – .	R . – .	4 . . . . –
G Golf	P Papa	Y Yankee	G – – .	S . . .	5 . . . . .
H Hotel	Q Quebec	Z Zulu	H	T –	6 – . . . .
I India	R Romeo		I . .	U . . –	7 – – . . .
Yes = Affirmative			J . – – –	V . . . –	8 – – – . .
No = Negative			K – . –	W . – –	9 – – – – .
			L . – . .	X – . . –	0 – – – – –

Table 1.1. *International radio signals.*

2.3.1.1 International radio signals

Radio reception varies according to the time of day, particularly in a tropical zone, where the sun is vertical. In the morning or evening better contacts can be made at low frequency between 2000 and 4000kHz, but for contacts during the day frequencies between 8000 and 12,000kHz must be used. The use of radio communications is also connected with safety questions. It is important to know:

1) how to send out a distress call;
2) how to contact a plane or guide it to the landing place;
3) what ground-to-air signal to use (**Fig. 1.2**).

2.3.2 DISTRESS CALLS

For a call by radio on the international distress call frequency, ie 500kHz every GMT half-hour at 15 minutes to and 15 minutes past the hour, there is a period of silence for 3 minutes.

First, call 'MAYDAY-MAYDAY' using voice and give your call sign and your position. If this fails, send an S.O.S. signal in Morse code, followed by 20 seconds of dash signals and repeat the call frequently for radio-direction-finding location. Another aircraft frequency is 2182kHz at 3 minutes past and 33 minutes past the hour.

Also obtain details of the local distress frequencies, the nearest airfield control tower, another user working in the same sector, police stations, forest wardens, etc. Have a distress frequency crystal if you have pre-set equipment.

2.3.3 MAKING CONTACT WITH A PLANE

When giving your position by radio to a plane, you can use a watch to establish your position, which you always give in relation to the pilot, ie in the direction the plane is flying, in hours (**Fig. 1.3**). You then give the angle of your position by radio, eg 'I'm at one o'clock in relation to you'.

If you do not have a radio, you can also use a mirror to establish contact with a plane (special survival mirrors are on sale in all shops which sell marine fittings).

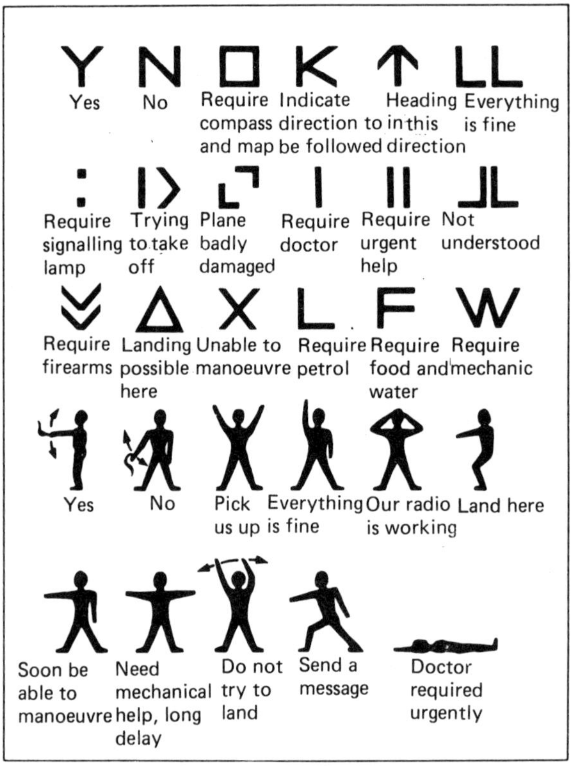

Fig. 1.2. *Ground-to-air signals used ICAO.*

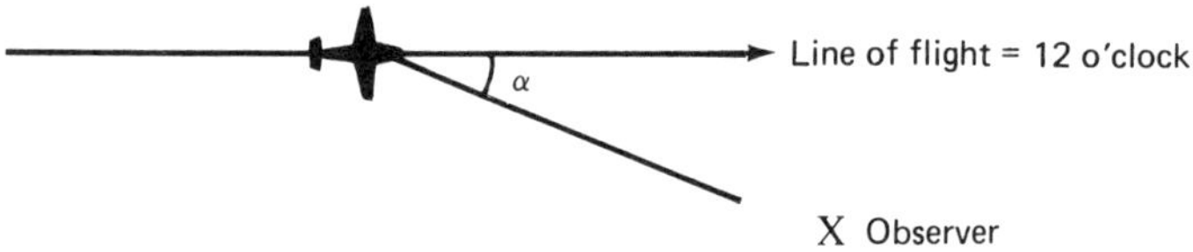

Fig. 1.3. *Observer's position for radio contact with a plane.*

Hold the mirror a few inches away from your face and aim at the plane through the hole (**Fig. 1.4**). The shaft of light formed by the hole will fall on your face, hand or shirt. Adjust the angle of the mirror until the reflection of the spot on the rear side of the mirror disappears through the hole while you are still looking at the plane through the hole. The reflection can be seen for up to 10km.

2.4 Orientation

In the northern hemisphere, outside the tropics, the sun is always towards the South but between the tropics it varies between 21 June and 21 December. Knowing the

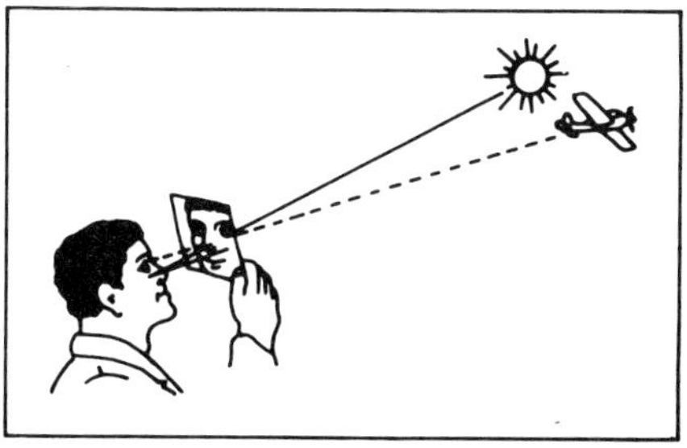

Fig. 1.4. *Using a mirror to establish contact with a plane.*

latitude and the date, you point the small hand of your watch, which shows actual local time, towards the sun. South is given by the line that bisects the angle formed by the small hand and the number 12 which depends upon the latitude. Conversely, in the southern hemisphere the sun indicates North **(Fig. 1.5)**.

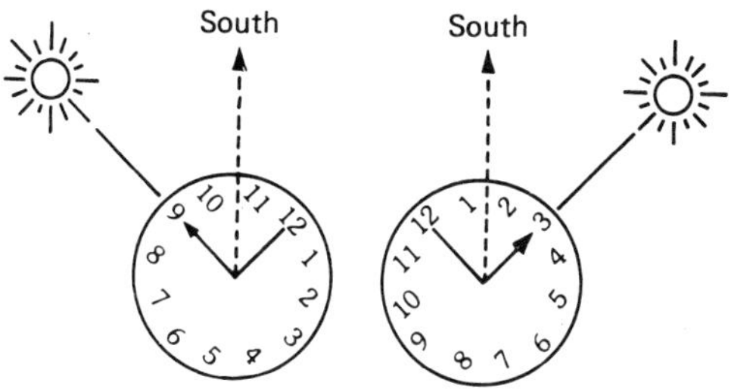

Fig. 1.5. *Orientation using a watch (northern hemisphere).*

Conversely, to tell the time using a compass you measure the angle between South and the sun. The same angle on the opposite side of the South needle gives the position of the 12 on the clock face and you read the time in terms of the angle between the imaginary 12 and the sun (Fig. 1.6).

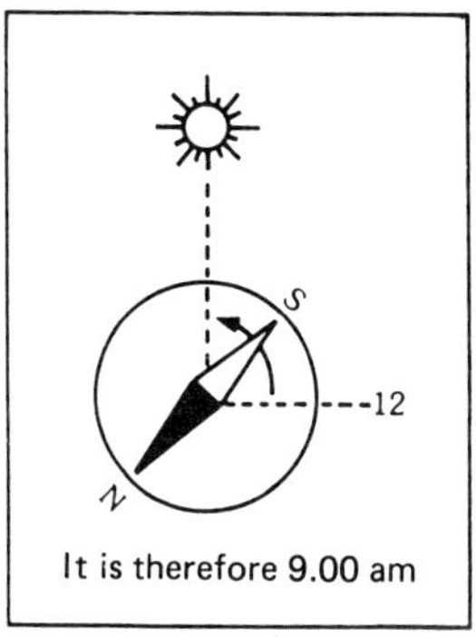

Fig. 1.6. *Telling the time with a compass.*

2.5 Rations

In certain countries if the working conditions require a long field trip, labour legislation requires that staff be fed. It is therefore advisable to enquire from the local authorities about the current legal rations in the country and to plan the supplies accordingly.

2.6 Expedition activities

2.6.1 CONTACTS ON ARRIVAL

After setting up the base camp at the chosen site, make contact with the various authorities, government, local administration, police, labour officer, post office, etc., to request their assistance. Recruiting staff is made easier if a representative from the local authority helps. Fix the daily wage according to the labour available and the current legal rate. The hours of work, rest time, overtime and the rate paid, transport and other allowances, porters loads (20kg), conditions of dismissal and notice must all be defined with the labour officer before the workforce is hired, in order to avoid any subsequent dispute. Theoretically, day labourers should be paid every day; but if they agree it is easier to pay them by the week or month or after field trips. In some countries, current legislation requires that the workforce be housed and equipped, so make enquiries beforehand.

In the absence of a medical test, which is in theory essential, make sure that the workers taken on have no obvious health problems (suppurating wounds, skin diseases, eye problems, coughs, etc.) (see the bibliography 'Medical Guide for Journeys').

2.6.2 ACTIVITIES AT THE CAMP

Setting the camp up well is never a waste of time, even for a temporary camp. The ground must always be cleared, with no vegetation, the toilets and rubbish pits far enough away so as not to annoy people and attract flies, rats, etc. The camp must also be safely situated. The local workers generally know where not to set up camp, because of animals, proximity to a water supply point, an area liable to flooding, etc.

2.6.2.1 Camp duties of the expedition or sector leader

These duties apply to the specific case of a strategic prospecting expedition:

1) programming and planning the project;
2) preparation of itineraries;
3) discussing itineraries, once there has been time to study them, with leaders of prospecting groups;
4) workmen's rations: calculating the quantities of food required for the month, planning stocks;
5) seeing to it that the condition of vehicles is checked and their maintenance regularly carried out;
6) checking fuel supplies, monitoring consumption;

7) supervising packing, listing and despatch of geochemical samples in accordance with established procedures, making provision for the necessary tubes and sample bags for the following month, etc.;
8) purchases of food, fuel, stationery, batteries, medical supplies, etc.;
9) administration: invoices, pay, recruitment, requesting cash funds for wages, etc.;
10) liaison with base;
11) preparation of maps for reporting results of work;
12) study of petrographic samples;
13) study of alluvial samples;
14) transfer of results to maps;
15) completion of a monthly report for submission on time;
16) study of points and sites of particular interest in the area;
17) distribution of rations to workmen.

2.6.2.2 Duties of the prospector and group leader
Records recorded on a 1:200,000 or 1:100,000 scale map as available:

1) precise route taken, if different from that planned;
2) location of petrographic samples;
3) location of alluvial samples;
4) location of geochemical samples.

Notebook record of:

1) the number of a geochemical sample and its relation with the alluvial samples;
2) description of alluvial samples examined through a binocular microscope;
3) description of petrographic samples.

Report of:

1) field work done daily, on index cards;
2) special observations, possibly on separate sheets.

Geological map of:

1) work with individual prospectors on petrographic samples;
2) selection of possible samples for thin section study;
3) selection of mineralized samples for polished section study;
4) study of particularly interesting alluvial samples;
5) transfer of field and petrographic data and the drawing up of the geological map on a 1:200,000 scale, or the scale being used.

By way of example, Appendix A3 contains some extracts from geological prospecting maps drawn up by the BRGM.

Name	No of alluvial samples taken	No of geochemical samples taken	No of petrographic samples taken	Kilometres covered	Field days
Thomson	25	32	34	41	7
Thomson Durand, etc.	45	142	10	114	14
Total	254	607	141	721	96

Table 1.2. *Statistics.*

2.6.3 MONTHLY REPORT

2.6.3.1 Technical report

1) Introduction: general observations on working conditions during the month, vehicles, illnesses, the stage that work has reached, etc.
2) Statistics: can be useful for general survey expeditions.

Senior officials attach importance to statistics, which do not necessarily prove the efficiency of the actual work done.

3) Geology: petrographic descriptions of geological units. When work on a particular subject is completed an appendix with sections, plans, maps, technical notes, etc. will be included.
4) Prospecting: observations after study of the alluvial concentrates, observations on any mineralization.
5) Conclusions: on the month's work and the progress made by the expedition.
6) Programme for the following month: extract from schedule, estimates; estimates of the number of alluvial and geochemical samples to be taken; route map to be included at the start of work in each sector, and to be modified each month; map, scale 1:1,000,000, showing the area covered during the month and the area to be covered during the following month.

Enclose two copies of each group's job files. N.B. Keep a spare copy of each report, including job files and technical notes or appendices enclosed.

2.6.3.2 Adminstrative report
The content of such a report varies according to the expedition and the region; take an example from the department's administration officer before departure.

3 Conclusions

This chapter describes the preparation for an expedition to a tropical area. Failure to forsee problems is far more serious in a tropical project which is in any case bound to be more hazardous and more costly than work in a temperate developed country. Therefore, much of the technical content of this chapter and the rest of the book is applicable to any project.

References

Advenier, W A; Pichot, R D (1975) Le guide médical des voyages. Collection Marabout service santé No. M S 254. Indispensable to any working geologist. Concerns tropical diseases in all countries: risks, methods of prevention, first aid in the absence of doctors, and symptoms to look out for.
Chaussier, J B (1980) Préparation matérielle d'une mission de prospection minière en zone intertropicale. Doc. BRGM, No. 23, 35 p. 18 fig.
Greenbank, A (1968) Comment sauver sa peau (à défaut de Survival). Ed. Stock. As the title indicates, survival techniques for all situations.
Marreua, B and C (1980) Le guide de tout-terrain. Ed. Brea, 24, avenue Ledru-Rollin, 75012 Paris. How to drive, equipment, choice of vehicle, characteristics, price, advice on all-surface driving, mud, sand, etc. Really intended for 'raiders' and all-surface rally drivers.
Survival, U S Air Force - AAF 64-01. Edit Air Technical Service Command, Wright Fields, Dayton, Ohio, USA. Survival packs for all western armies, as essential as compass and knife. Aid to survival at sea, on land, in desert, jungle, arctic. If you only have room for one book, make it this one.

Appendix 1

Equipment for the expedition

The following lists are given simply as guidelines for choosing equipment. The selections actually made will depend on the length and the type of the expedition (eg a permanent camp for twenty or more people, or a travelling camp for one person).

1 Camping equipment

This list was drawn up with a general prospecting expedition (eg of a BRGM survey in an equatorial forest) in mind.

Two-man tent with a spare fly-sheet
Light-weight tents for fly-camping with spare fly-sheets
Ropes for tent stays
Tarpaulins, 5 × 4m
Tarpaulins, 2 × 2m
Light-weight tent for travelling
Folding lounging chairs with spare parts
Folding chairs
Light-weight folding stools
Folding tables

Roll of canvas for repairing chairs, etc.
Folding beds with spare parts
Light-weight camp beds
Bedding trunk and padlocks
Sleeping bags
Sleeping bag sheets
Mosquito nets
Sheets
Blankets
Pillows
Pillow cases
Matches or lighters
Gas cookers (for base camp)
Portable gas stoves
Gas carboys
Gas regulators and tubes
Portable kerosene (paraffin) stove (not absolutely necessary)
Refrigerator (electric, kerosene or gas)
Freezer for base camp (electric, kerosene or gas)
Cleaning brush for kerosene fridge flue
Spare glass for kerosene fridge
Spare glass for kerosene freezer
Spare wicks for kerosene fridge and freezer
Kerosene filter funnel for fridge
Portable ice boxes, large and small
Blocks of Coleman dry ice
Small refrigerator powered by a car battery
Camping showers (water bag type) or shower buckets
Metal shower tub
Metal buckets
Plastic buckets
Plastic bowls
Plastic jerrycans (20 litre/4 gallon) for water
Plastic jerrycans (10 litre/2 gallon) for water
Plastic jerrycans (5 litre/1 gallon) for water or kerosene
Canvas water bags
Torches powered by batteries
Spare batteries and bulbs
Water filter with accessories and spares
Manganite kit (if necessary)
Spare manganite
Lamps with mantles, spare glasses and cartridges, if the small type is being used
Kerosene lamps (eg Coleman, Tilley, Petromax) with mantles, spare glasses, etc.
Hurricane lamps with spare wicks and glasses
Packets of candles
Empty trunks and combination padlocks
Tool kit for general repairs

2 *Cooking equipment*

Frying pans, large and small
Pyrex dishes with lids, large and small, for roasts
Casserole dish
Pressure cooker
Grill pan
Saucepans and lids
Plates, large and small
Mugs, cups and saucers
Soup and dessert dishes
Salad bowl
Carving knife and fork
Bread knife
Knives, forks and spoons
Cooking spoon, fish slice, ladle, etc.
Glasses or beakers
Vacuum ('Thermos') flasks
Colander
Funnels, small and large
Vegetable grinder
A large number of airtight plastic containers for sugar, rice, flour, etc.
Tin opener
Bottle opener
Sardine tin key
Corkscrew
Electric iron (if necessary)
Flat iron (for the bush)
Teapot
Coffee pot
Kettle
Tea towels, towels, tablecloths, napkins
Washing-up liquid
Sponges, scouring pads
Washing powder, soap

3 *Prospecting and camping equipment*

Geologist's hammmers
Universal compasses in cases
Map case or board with clips
Field notebook case
Rucksack with pockets
2m tape measures, pocket size
Pocket stereoscope
Pocket altimeter
Pedometer

Magnifying lenses for 8 or 10× enlargement
Plane table
Hydrochloric acid in a plastic container
Steel cables
15mm nylon ropes
5mm nylon cord
Hand-winching device
Hacksaws, pliers
Petrol chainsaw and three chains
Round file for sharpening
Picks and shovels
Hammer and wood chisels
Pincers, pliers, plane, hand saw, axes
Machete (one per labourer)
Files (one for every five labourers)
Bradawl or spike
Sledgehammers (3 and 5kg)
Chisel, crowbar
Drilling rods (for flat prospecting)
60cm and 120cm drill rods and extension pieces
Breakout tongs
Pulleys for prospecting pits
Pumps with pipes
Protective helmets, possibly with lights
Ordinary rucksack
Jute bags
Spare 'topofils' (see Chapter 2, Section 1.1.3) and reels
10m reel-tape in plasticized cloth
Desk stereoscope
Binocular microscope
UGINE magnets for the camp and a small number for the field (pencil type)
Box for Mohs scale of hardness
Assay balance and weights
Thin steel cables for pits
Various fish hooks and nylon line for fishing
Weapons and ammunition (possibly)
Hand crusher
Steel mortar and pestle
Porcelain mortar and pestle
Brushes for cleaning mortar
Dividers, large and small with tub and scoop
3mm prospecting sieve
8 litre and 2 litre pans
Wash pans (chinese hats)

Plastic bags for sediment samples
Series of sieves, diameters 500, 315, 160, 125, 80μm with bases and covers
Basins, metal buckets
Plastic sample tubes large and small
Plastic bags, about 70 × 40cm
Cloth sample bags for rock samples
Scintillometer or Geiger counter (see Chapter 5)
UV lamp, short wave, long wave
Spare batteries for equipment
Yellow labels (plastic)
Cardboard labels for tubes
Aluminium labels for staking in the field and marking the base points and reference points for tactical prospecting

Geochemical reagents
Zinc sheet
Lead sheet
Porcelain sheet

Portable radio
Dry cell or car battery with spare fuses
30W battery/d.c. radio for base camp with spare fuses

Generator set
Spare fuses and wire
Spark plug sockets
Spark plugs
Electric cables, sockets and lamps

4 *Office equipment*

Calculator and spare batteries
Portable typewriter
Typewriter paper
Typewriter copy paper
Carbon paper
Typewriter ribbons
Envelopes, small and large
Tracing paper pads
Graph paper pads
Files with straps
Loose-leaf files
Coloured folders, paper and card
Writing pads, field notebooks

Sample notebook
Drilling notebook
Large and small format duplicate notebooks
Small stapler and staples
Large stapler for sample bags and staples
Paper punch, 3-pointed drawing pins, paper clips, adhesive tape, inking pad, string, glue
Kodatrace and wax pencils for drawing on aerial photographs
Rolls of tracing paper
Pens and felt-tip pens, assorted colours
Scissors
Lead pencils, HB and H, coloured pencils
Marker pens for labelling sample boxes
Erasers, typewriter erasers, pencil sharpeners
Rulers, set squares, protractors
Drawing instruments
Rapidograph (0.2-0.3-0.5)
Drawing ink
Drawing curves
Drawing board
Tubes for maps and tracing paper

Equipment for administrative work
Metal cash box, accounts book, pay books, cashbooks, attendance books, etc. and all necessary administrative documents according to the country and employer
Postage stamps

5 First aid chest

Malaria tablets
Soluble aspirin
Vitamin C, tonic
Salt tablets
Multivitamin pills, to supplement poor diet
Intestinal sulphamine against enteritis
Intestinal dressing for enteritis
Antispasmodic
Cough mixture
Ointment for muscle strain
Eye lotion
Ear drops
Throat lozenges for tonsilitis
Antibiotic tablets

Antibiotic ointment for external use
Insect repellant
Pain-killing ointment for insect bites
Snakebite serum (two sets per group) and cardiac stimulant for administration after injection of serum
Cotton, gauze, biogauze bandages for burns
Disinfectant for wounds
Mercurochrome for drying wounds
Sulpha powder for infected injuries
Talcum powder
Scissors, fine tweezers for splinters
Scalpel, suturing needle
Butterfly clips for closing wounds
Clamping tweezers, tourniquet, thermometer (several as they are so easily broken)

It is advisable to obtain a first aid manual before departure, to equip oneself with normal everyday medicines and to consult a doctor (see References).

6 *Equipment for the vehicle*

To be adapted to the type and make of vehicle

Tools
Hydraulic jack with crank (or lever)
Spare wheels
Wheel spanner
Tyre levers
Foot pump
Grease gun
Padlocks
Wire brush
Tool kit including:
- adjustable monkey wrench
- set of spanners, Imperial or metric sizes
- adjustable spanner
- box spanner
- Allen key
- multigrip pliers
- pliers
- hinged spark-plug socket
- pressure gauge
- mechanic's hammer
- chisel, pin driver
- set of flat and cross-head screwdrivers
- spark plug brush

20 litre/4 gallon metal jerrycans for petrol
10 litre/2 gallon plastic jerrycans with pouring lip for water
1.50m plastic pipe, 1 to 2cm diameter, for use as petrol siphon
Nylon tube, 7 to 10mm diameter, 25cm long for the brake
Machete, hand axe
2kg sledgehammer
Shovel, pickaxe, crowbar
Towline, 1.5 to 2cm diameter, length 10 to 15m, with eyes or hooks at the ends of the line
12V handlamp, which can run off the battery
Battery clips
Starting handle
Oil can
Pair of gloves
Spare parts
Puncture repair kit
Complete front and rear springs
Upper and lower radiator hoses with clamps
Fan belts
Mechanical petrol pump
12V coil
Distributor cap, rotor arm and points
Condenser
Spark plugs and rubber spark plug protectors
Bulbs for dipped headlights, sidelights and indicator lights
Assorted fuses
Roll of insulating tape
Oil filter
Pair of spring shackles
Spring bolts with nuts
Spring counter bolts
Front half axle (half-shaft)
Rear half axles, left and right
2 litre/0.5 gallon can of oil
Brake fluid
Fire extinguisher
Bottle of distilled water
Assorted joints/gaskets

Vehicle maintenance
See vehicle instructions
Wash before all operations
Motor oil change every 2000km (clean the air filters often)
Gearbox
Front and rear axles
Differential
Oil filter change every 6000km
Grease-points

Brake adjustment every week
Periodic checking of tyre pressures
Welding kit for repairs

7 *Equipment for boats*

Outboard motor (5, 10, 18, 20 or 40HP, according to the size of the boat)
Fuel tank
Fuel supply tube, tank to motor
Funnel for fuel
Transparent nylon fuel filler hose 1.50m long, 2.5 to 3mm diameter
Paddles, wooden poles
Bucket for baking
Coil of rope
Mooring anchor
Nylon towing rope, 1.5 to 2cm diameter, 50m long
5m mooring chain with padlock(s) (preferably combination)

At the camp
Stand for motor when not in use and an empty petrol drum filled with water for testing the motor on the stand

Tools
Hammer
Screwdrivers of various diameters
Cross-head screwdriver ('Pozidriv', etc.)
Spark plug socket
Small grease gun
Spark plug brush
Small pocket knife
Multigrip pliers
Shifting spanner
Pin driver/punch
Plastic bowl for measuring oil for oil-fuel mixture
Syringe for injecting oil

Spare parts
Kit of spare pins/bolts for propellor axle
Spark plugs
Rags/cloths
Jerrycan for extra fuel

8 *Personal kit*

The kit must be suited to the climate, be as light and as manageable as possible and must not exceed your carrying capacity, ie 5 to 8kg, taking into account that you also have a compass, hammer and a lunch bag for carrying maps and a flask. The following are recommended:

a) Sun glasses.
b) A hat which is as light as possible (denim perhaps, certainly not a pith helmet, which is far too heavy) but which covers the nape of the neck, the ears and the eyes, and provides protection against both sun and rain. Bush hats are too heavy, making one perspire.
c) Cotton underwear which avoids chafing from perspiration and the likelihood of fungal infections.
d) Ordinary light but tough (perhaps denim) clothing.
e) Practical overgarments suited to the climate. Even in a hot climate a quilted jacket can be useful in the evening or early morning; a light waterproof gives protection against both rain and wind. Clothes may need to be tough enough to withstand thorns; an army surplus combat jacket is ideal.
f) Above all, whatever the climate, never wear clothes which are too tight. Remember that the clothes worn by the people of the country concerned are always best suited to the climate.
g) Footwear should be of a kind that grips the ankle well, does not make the wearer's feet hot and allows him to walk through water, the water draining through built-in holes to ensure quick drying. In desert or semi-desert regions the leather boots worn locally are preferable because more air reaches the feet; they also protect against mosquitoes, scorpions, snakes, etc. To avoid picking up parasites, such as roundworms, hookworms and bilharzia, never go barefoot.

Drink only filtered water to avoid amoebic dysentery, cholera and typhoid, and to avoid bilharzia do not bathe in non-running water.

Make sure that mosquito nets are intact. Always have some snakebite serum with you and if you cannot keep it in a efrigerator, renew your supplies as often as possible. Take an antimalarial prophylactic regularly and continue doing so for two months after returning. Also remember that every prospector must always carry a packet of tissues, a multipurpose knife and an emergency whistle.

Camping out overnight, remember to shake your clothes before dressing and to check that no insects are hiding in your boots.

9 *Hunting/Fishing*

Hunting ensures a supply of fresh meat so long as it is allowed. Normally a good local hunter is asked to handle this job. Take fish hooks and some nylon fishing line for the labourers when visiting a region where fishing is enjoyed.

10 *Diet*

Take a cookery book with recipes that can be adapted to local products and will give a balanced diet. Some overnight camping field trips may need concentrated 'survival rations' that can be obtained from specialized shops or from the diet sections of large department stores. Failing that, make up a ration kit yourself with peanuts or peanut butter, sweetened dried fruit, concentrated milk, brewers' yeast and salt and multivitamin tablets.

Chapter 2

Hammer prospecting

'Hammer prospecting' is the search for indications of mineralization by the observation of outcrops, debris or 'float', etc. and is made along routes or traverses planned from the documentation available, geological, geochemical and geophysical maps, photogeological maps, maps showing outcrops and prospects, mineralization indexes, etc., and after consulting the local population. Little equipment is needed for the initial search. Heavier equipment, such as light drills, small motorized drills and excavators, is only necessary at a later stage for the study of the discovered individual prospects.

1 Tools, instruments and reagents

Hammer prospecting requires the use of a number of tools and instruments whose handling presents no special problems. Local conditions or the search for a particular kind of mineralization may require additional material, and a choice has to be made according to the type of work.

1.1 Description and use of instruments

An instrument is always provided with instructions for its use but nevertheless there are a few points worth stressing.

1.1.1 THE COMPASS

Various types of compass exist but the most common in the BRGM is the Chaix universal compass, which can be used for both topographical and geological surveys.

1.1.1.1 Topographical surveying
The observer is a fixed point from which sightings are made, all of which therefore have a clearly defined direction. The azimuth is measured by reading the number shown in the magnifying prism. The clinometer can also be used to measure angles of slopes approximately, by taking a bearing, with the sighting wire parallel to the direction of maximum slope on an object of the same height as the operator. The compass is then aligned vertically, and the clinometer needle is released to measure the slope. The true horizontal distance is found by applying a correction coefficient

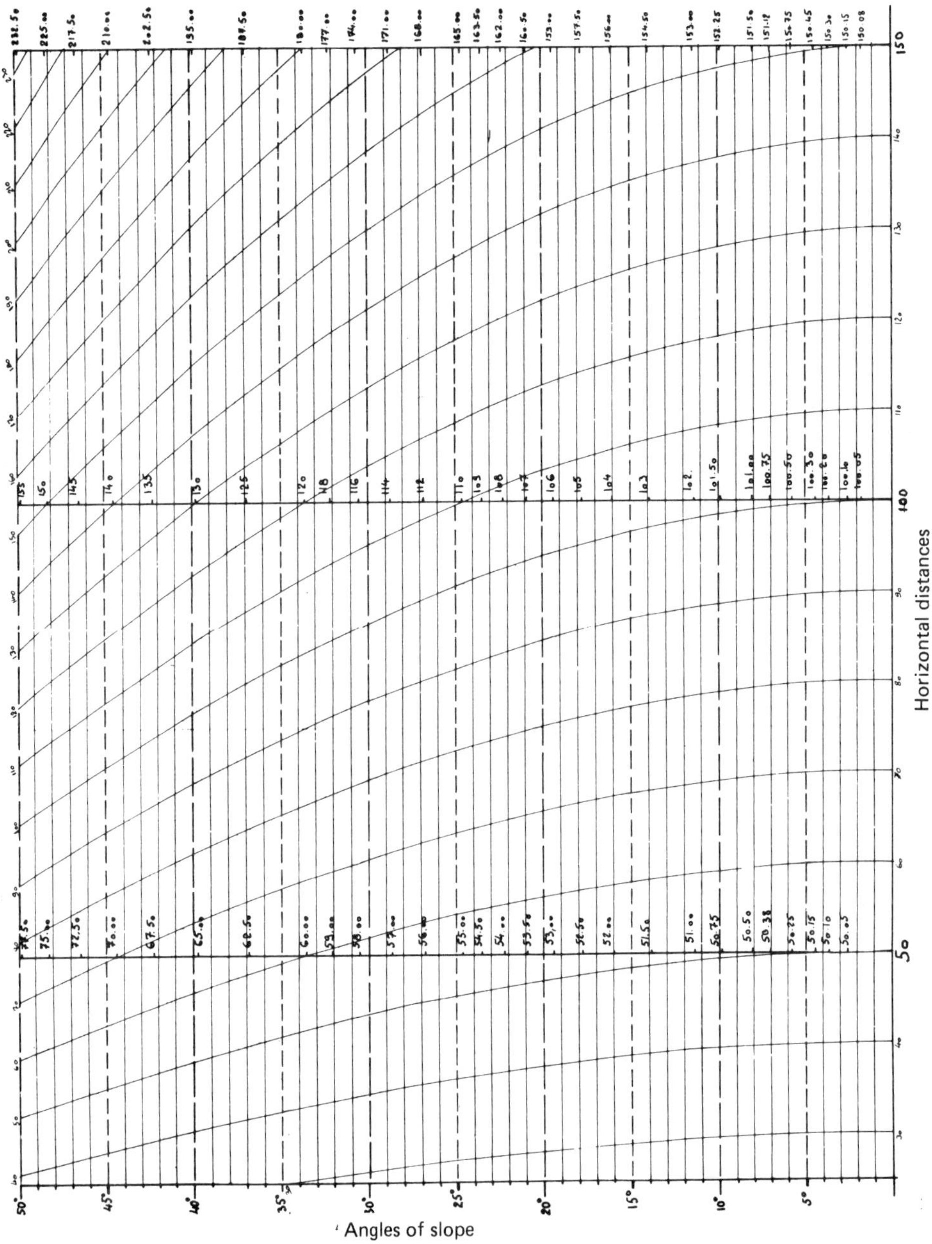

Fig. 2.1. *Slope correction.*

K to the distance measured along the slope. **Figure 2.1** shows a chart from which the correction can be read.

1.1.1.2 Geological surveying

Measurements taken include the strike and dip of veins, bedding, schistosity and fractures, as well as the direction and plunge of fold axes, lineations and ore shoots, etc.

Directions

Directions are not recorded in quadrants (eg N30E, S50W) but are read directly in degrees from the compass. This simplifies plotting on a Wulff net and computer storage of data. On the compass the angle increases in a clockwise direction, so that N30W becomes 330° (**Fig. 2.2**).

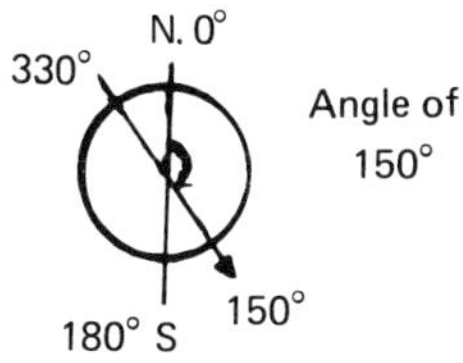

Fig. 2.2. *Compass angles.*

Magnetic North and true North

All compass readings are relative to Magnetic North, which varies in space and time. It is therefore important, for maps made by compass survey, to give not only geographical coordinates but also the month and year when the survey was made. Magnetic North should be shown even if the declination (the angle, East or West, between True and Magnetic North) is known. National geographical or geodesic organizations can provide the declination at any given point.

Some compasses have an adjustable graduated ring that can be turned from Magnetic North, previously aligned with 0° on the ring, through the angle of declination, so that readings made are instead related directly to True North. The graduated ring is turned to the right (clockwise) or left (anticlockwise) depending on whether the declination is an East or West value. This modification is useful when the prospector has a topographic base on which to plot his geological observations directly as the precision required is no more than half a degree or even a degree. In cases where he has to make his own topographical map by compass survey, however, it is preferable to leave the adjustable ring oriented to Magnetic North. This fact should be noted on the map, together with the date.

Dip

The dip of a bed is defined as the angle between the horizontal and a line drawn perpendicular to the strike in the bedding plane. Certain precautions must be taken when measuring dip.

If the dip is greater than about 30°, the line of maximum inclination is easy to locate but at lower values the following steps should be taken:

1) Locate the strike of the plane of stratification, ie the direction of a horizontal line in the plane. First press the button that releases the clinometer needle. Keeping the edge of the compass on the bedding surface

and the compass vertical, turn the compass until the clinometer reads 0°. The line now marked by the edge of the compass on the bedding plane is horizontal, and corresponds to the strike.

2) Keeping the edge of the compass in the same direction, turn the compass into the horizontal position and read the bearing.
3) Now place the side of the compass against the bedding plane, aligned perpendicular to the strike. With practice this can be done by eye. Press the clinometer release button, wait until the needle is steady, then clamp it again. The angle now indicated is the dip (**Fig. 2.3**).

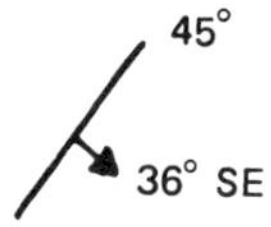

Fig. 2.3. *Dip.*

Because of the increasing importance of structural analysis, strikes and dips should be recorded to the nearest degree if they are to be treated statistically. They can be recorded in a field notebook as follows: Strike 45°, dip 36° SE or more simply keeping the direction of dip always towards the observer's right 36/45.

1.1.2 THE CLINOMETER

This device is seldom used now and has been effectively replaced by the Topochaix universal compass clinometer or by other combined clinometer/compasses (Brunton).

1.1.2.1 Dip measurement

Place the upper or lower edge of the compass on the geological layer. Press the button thus freeing the pendulum, then release it. When the needle stops moving it will give the dip in grades, degrees or %.

1.1.3 'TOPOFIL'

The 'topofil' is a measuring instrument which has a counter to record the length of line coming out of the instrument when the user travels the distance being measured. The counter can be accurate to the nearest decimetre or metre; for hammer prospecting, measurements to the nearest metre are sufficient. The line is 5000m long and is wound on a spool; there are two spools in the instrument and therefore distances of up to 10km can be measured.

When using the 'topofil', two things should be borne in mind:

1) in a car the measurement is only accurate if the car is travelling at less than 15km/h;
2) long distances (over 30m) can be distorted by the effect of strong wind on the line. In this case it is advisable to attach the line to branches or posts.

Distances can also be measured with a speedometer consisting of a wheel connected to a kilometre counter.

1.1.4 CHAIX TOPOGRAPHICAL BOARD

The topographical board is a portable instrument which allows results from ground surveys to be written out without difficulty, ie it records given distances following given directions. This is recorded on tracing paper which is fixed on to two rollers, so that the paper can be moved across the board. The directions are marked out using a 0 to 360° circular protractor which turns on its central axis. Across the diameter of this circle there is a groove, graded in millimetres, which marks out distances being measured. The protractor is fixed on to a carriage which can be moved sideways.

1.1.5 OTHER INSTRUMENTS

For other detailed operations requiring a high degree of accuracy a board placed on the ground with a clinometer ruler should be used, and the instruction manual for the equipment should be consulted for advice on its use.

1.2 Use of reagents

Chemical analysis work in the field requires certain instruments and products which make a fixed base camp necessary. However, a few reagents which can be handled easily will enhance observations made in the field.

1.2.1 PREPARATION OF SAMPLES

If a reagent is to produce a clearly visible colour, the sample must be carefully prepared, in particular to eliminate excessively fine particles or dust which would cling to the grains being observed and prevent proper colouring.

Dust particles can be removed by washing the sample in running water, or when working on a fixed rock, a freshly fractured surface should be used for analysis.

1.2.1.1 Limestone identification using a 10% solution of hydrochloric acid

1.2.1.2 Qualitative determination of zinc ('Gertrude' reagent)

This reaction is based on the ability of alkaline ferrocyanide to oxidize certain aromatic amines, especially diethylaniline. The reaction is very slow except in the presence of zinc, since ferrocyanide is coloured by adsorption of the resulting organic product.

To be prepared:

1) a solution of potassium ferrocyanide (3%) – $K_6Fe(CN)_6$;
2) a solution of oxalic acid (3%) in which 0.5% diethylaniline is dissolved.

The quantities to be prepared vary according to the length of the field expedition.

For long expeditions (some months) several litres of each solution should be prepared and kept at the base camp. Flasks of solutions containing diethylaniline should be covered with aluminium foil, since this solution deteriorates when exposed to light.

'Gertrude' reagent is a mixture of these two solutions (equal quantities of each). It should be made up the day it is to be used, since it deteriorates after two days, and preferably kept in a non-metallic, unbreakable opaque pocket-sized vaporizer (plastic flask). It is then sprayed on to the outcrop or the sample. A blood-red colour will appear after a few seconds or minutes. This reaction will occur with smithsonite, calamine and hydrozincite.

In the case of blende (sphalerite) the same results can be obtained by superficially carbonating the sample with a solution of sodium hypochlorite (domestic bleach) containing 10% sodium carbonate for about ten minutes. The sample should then be washed thoroughly and the same reaction will occur.

1.2.1.3 Qualitative determination of lead ('Jerome' reagent)

This reaction is based on the precipitation of lead iodide which is yellowish-gold colour.

To be prepared:

1) an acetic acid solution (5%) (or 10% hydrochloric acid);
2) a potassium iodide solution (5%).

The quantities prepared vary according to the length of the field study. For long expeditions (some months) several litres of each solution should be prepared.

'Jerome' reagent is a mixture of these two solutions (equal quantities of each). It should be made up the day it is to be used, since it deteriorates after two days, and kept in a vaporizer. It is then sprayed on to the outcrop or sample. If lead is present, a characteristic yellow colour will appear.

1.2.1.4 Qualitative determination of copper

The reagent used to determine the presence of copper in dark shale is a sulphide reagent but the reaction is not specific to copper alone. Interpretation of results, therefore, requires care and experience.

A brush should be used to paint a narrow white strip (the make-up of the paint is described below) on the sample. If chalcocite copper is present, a blue colour will quickly appear. With other sulphides colouring will be more gradual, occurring after several seconds (pyrite, galena, blende).

1) Preparing the paint: 40g of ammonium molybdate and 20g of ammonium pyrophosphate are dissolved in 10cl of distilled water.
2) A solution of 170g of titanium dioxide and 10cl of concentrated hydrochloric acid is then added to give the liquid the viscosity of paint.

1.2.1.5 Testing for copper on neotocite

Cupriferous neotocite forms a chocolate brown coating on fractures. This can be scraped off as a powder using a nail. Some drops of HCl (10% dilute) are added, into which the nail is dipped. The result is positive if copper forms on the nail.

1.2.1.6 Qualitative determination of antimonic minerals
This reaction is based on the precipitation of an orangey-red salt. Two solutions are used (A and B, see below), the second applied one minute after the mineral is attacked with the first. If antimony is present, an orangey-red colour will appear. (As and Bi can disrupt this by also producing colour, although less marked.)

1) Attacking solution A = 1 vol concentrated HCl, 3 vols concentrated H_2SO_4 and 2 vols distilled H_2O.
2) Reagent solution B = 10g potassium iodide, 2g ascorbic acid, 40g caesium chloride dissolved in 10cl of distilled water.

1.2.1.7 Qualitative determination of phosphate
Two flasks are needed, one containing 50% dilute nitric acid and the other 12.5g of ammonium molybdate in powder form mixed with 100cc of water. The rock should be scrubbed clean using the nitric acid and sprayed with the solution in the second flask. If the rock contains phosphate, a bright yellow colour will appear.

1.2.1.8 Qualitative determination of tin minerals (cassiterite)
Grain samples are placed on a zinc metallic plate with its sides turned up and concentrated hydrochloric acid is added. This is then left for about 10 minutes. A thin layer of metallic tin forms on the cassiterite grains. This can be seen by eye (on dark cassiterites) or under a microscope or magnifying glass (on light cassiterites).

2 A few things to remember

It is important to distinguish:

1) Anomalies, which may be geochemical or geophysical and are the first indication of mineralization. They may be invisible at the surface, or visible in the form of compressed or trace mineralizations.
2) Traces, showings or indications, which give some idea of the degree of mineralization.
3) Deposits or ore bodies, which indicate that there are sufficient quantities available to be mined.

The contact planes between a mineralization and the country rocks are called walls. The upper wall is known as the hanging wall and the lower wall is known as the foot wall. The thickness of a mineralization is its breadth measured perpendicularly to the walls. The breadth measured horizontally is known as the horizontal crossing **(Fig. 2.4)**.

2.1 Surface weathering of deposits, gossan

The mineralization of an ore body outcrop is almost always different from that which makes up the greater part of the deep ore body. This difference results from oxidation accompanied by more or less complete disappearance of some elements

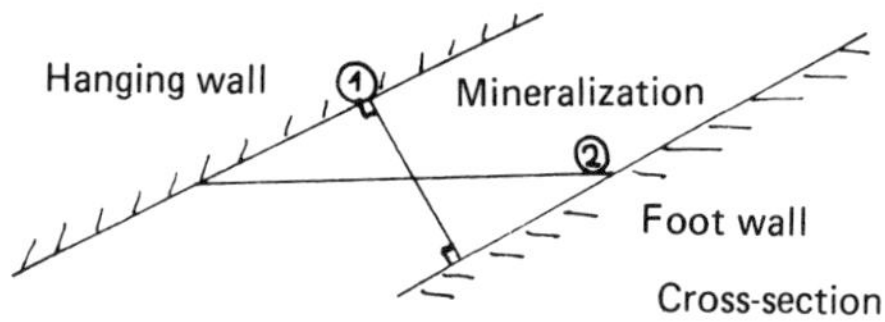

Fig. 2.4. *Mineralization cross-section: 1, thickness; 2 horizontal crossing.*

by leaching and from the concentration of other elements. This oxidized coat is known as 'gossan'. As they disappear, the crystals of valuable minerals or gangues sometimes leave characteristic cavities behind them, often covered with secondary

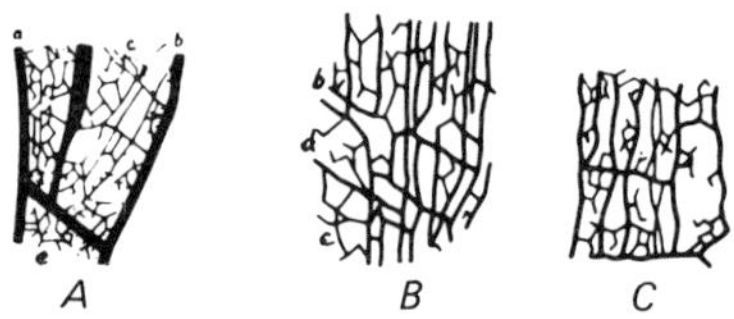

Coarse cellular, chalcopyrite boxworks

A × 2 *a*) quartz veinule; *b*) cellular walls of limonitic jaspar; *c, e*) interstitial limonite

B × 4 quadrangular appearance with parallel cellular walls; *b*) cross structure; *c,d*) fine, angular network

C ×5 average type

Coarse, cellular blende boxworks × 2 *a* quartz veinules *b* cellular walls of limonitic jasper; *c* interstitial limonite

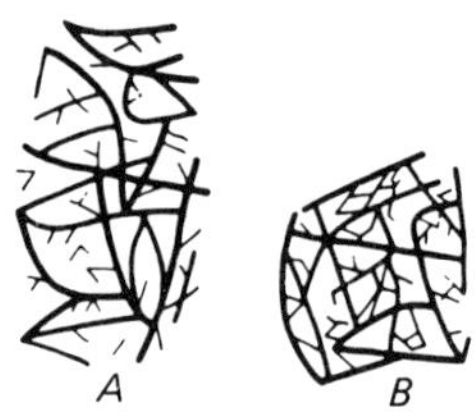

Triangular, curved bornite boxwork with eye-like cellules X 5

Foliaceous molybdenite boxwork containing limonite flows and quartz veinules

Curving tetraedrite boxworks (comparable with level curving) X 5

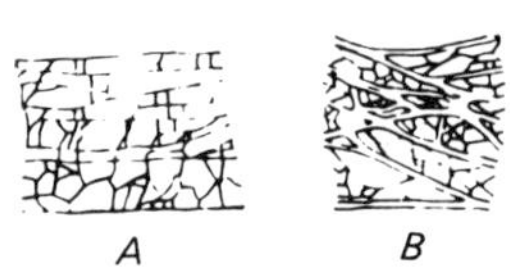

Galena boxworks

A X 3 standard appearance

B X 3 diamond mesh type of very fine galena

Fig. 2.5. *Types of boxwork.*

minerals produced by weathering of the main mineral. These cavities, known as 'boxworks', sometimes make it possible to determine the nature of mineralization of the main part of the non-oxidized part of the ore body but it is important to remember that the characteristics of boxworks often vary locally and that only a meticulous examination of the minerals to be found can allow precise determination of the nature of mineralization.

Table 2.1 and **Fig. 2.5** give extracts from work carried out by A M Bateman on boxworks.

Table 2.1. *Types of boxwork in limonite.*

Characteristics
Crude, angular; thin walls, wide, rigid; bubbles, masses
Siliceous, thin, rigid, angular walls
Thin, small, friable walls, flecks, bubbles
Stronger cells; 'banded' limonitic jasper
Rounded, thick, rigid, empty cells; fuzzy, a lot of silica
Triangular cells; thick, fragile, encrusted
Triangular, incurving
Long, close together, angular, rigid cells
No boxwork; arrangement of sulphide grains; fragile, powdery, in relief
Pitch-like; glossy, no cells
Thin, fragile, concentric lamellar sheets
Thin parallel cubic planes of limonitic jasper
Diamond-shaped mesh
Step-like arrangement
Smooth, thin, rounded cells

Boxwork	Fig	Colour	Deriving from
Crude cellular	3.1	Ochreous	Chalcopyrite
Crude cellular	3.2	Light brown	Sphalerite
Fine cellular	. . .	Orangey yellow	Bornite-chalcopyrite
Fine cellular	. . .	Light brown	Sphalerite
Cellular sponge	3.8	Brownish	Sphalerite
Triangular	3.9	Orangey ochre	Bornite
Triangular	3.3		
Curved	3.5	Orangey ochre Chocolate	Bornite Tetrahedrite
In relief	. . .	Chestnut	Chalcocite Covellite Bornite
Limonite pitch	. . .	Dark brown	Chalcopyrite
Limonite crusts	. . .	Dark brown to black	Chalcocite
Cleavages	3.7	Orangey ochre	Galena
Diamond mesh	3.6	Orangey ochre	Galena
Pyramidal	. . .	Orangey ochre	Galena
Foliaceous	3.4	Tan to chestnut	Molybdenite

Gossans form where there is relatively large-scale mineralization. On the other hand, where minerals are spread out and contain pyrite, weathering of the latter leads to bleaching and an often spectacular reddening of the country rock.

2.2 Alteration of ores

Alteration is a very useful guide for mineral prospecting. Numerous deposits are accompanied by an alteration surface which leads to a change in the colour of the rocks (often they are lighter shades, even white), in their porosity (leached altered rocks, often porous and light-weight) and in their mineralogical composition: sericitization, chloritization, silicification, pyritization, carbonatization, kaoliniza-tion, propylitization (appearance of a mixture of chlorite, amphiboles, carbonates and epidote). Weathering was discussed in Section 2.1. Alteration is the term used throughout this book to describe primary (metasomatic, hydrothermal) weathering.

3 Organization of the work

It is not uncommon to carry out prospecting in an absolutely unknown territory. Usually at least the beginnings of a geological map with indications of interesting mineral substances marked on it are available. The first phase of evaluation of a region, which can be classified as the exploratory phase, is very often done by national geological surveys. The basic maps are usually drawn to a scale of 1:500,000 or 1:200,000 (or 1 degree squared paper for intertropical zones).

Prospecting a predetermined region is therefore determined by the results obtained at the time of exploration. Two different cases may present:

Case 1: indications were found at the exploratory stage but by analogy with neighbouring regions the geological and structural conditions suggest the presence of valuable minerals. In this case the aim of general prospecting is to obtain indications, the importance of which will be confirmed as they are discovered. The scale adopted is usually 1:100,000 or 1:50,000.

Case 2: indications were found at the time of exploration or while searching for them. The general prospecting programme should thus be organized according to the relationship of these indications to the geological formations and the tectonics of the region. The aims of prospecting are to confirm the accuracy of the indications to examine their extent and, according to the data already obtained, find others. The scales adopted are usually from 1:20,000 to 1:1000.

In each case the work organization will differ slightly but all observations should be carefully noted when the survey is written up and shown on itinerary maps and maps of outcrops, samples and indications. In certain regions where outcrop conditions are bad the observations can be recorded on just one or two maps.

3.1 Exploration for indications

The aim of this exploration is to identify areas to be studied in more detail in the final phase or to stop all exploration in the prospected area.

3.1.1 TOPOGRAPHICAL SURVEY OF TRAVERSES

Where possible, the traverses can be directly marked on a topographical map, an aerial photograph or a drainage map drawn from aerial photographs. In tropical forests a map of the drainage systems is indispensable, since outcrops are often found near rivers. Where no documentation exists, or if the aerial photographs show mostly dense forest, the topographical survey of the traverse should be carried out as follows:

1) Take compass readings for lengths of 20m or more measured with the 'topofil' (if the landscape is hilly, make slope corrections using a compass/clinometer).
2) Place stakes at 100m intervals. Mark these stakes with a traverse marker (eg A, B, C) and with the distance in hundreds of metres from the departure point using coloured chalks (eg 1, 2, 3) or felt-tip pens or knock metal stakes bearing small numbered discs into the ground (this system allows the spots to be found long after the expedition is over).
3) Directions are recorded either straight away in the field or later at the camp if weather conditions are poor or if the compass alone is insufficient (then transferred into a notebook). They are recorded on tracing paper using a topographical board and then the traverse is marked on the map; if a small scale is being used, it is often necessary to put the results first on to a large scale map which can then be reduced photographically or using a pantograph.

Obviously this method is only necessary for winding tracks (slopes, streams, etc.), otherwise the itinerary can be simply traced on the ground by aligning three stakes and using stakes for slope correction. It is essential to use numbered stakes, since it is often necessary to go back to an earlier station. Moreover, in a dense forest area the tracks are open and are surveyed by groups of four men: one man with a compass is in charge of the group, one man has the 'topofil' and the others clear away undergrowth using machetes or curved knives. It is only several days later that the prospector or geologist arrives to begin prospecting.

3.1.2 THE USE OF THE SCINTILLOMETER AND UV LAMP

Uranium prospecting has become so important that it is recommended that a scintillometer or a gammameter with a high-sounding signal be taken on all expeditions. The scintillometer can equally well be used for phosphate prospecting (it detects the presence of U phosphates) and beach sands (zircon). Where there is a possible presence for fluorescent minerals, such as scheelite in the skarns or uranium-bearing minerals, an UV lamp should be used. This can be used also in the field at night for sampling outcrops, soils or alluvium (see **Table 2.2** for a list of the principal minerals fluorescent under UV light).

Table 2.2. *Principal minerals fluorescent in ultra-violet light.*

Minerals	Short wave length U.V.	Long wave length U.V.
Autunite	greenish-yellow	greenish-yellow
Anglesite	yellow, pink, white	yellow, red
Ankerite	orange	red
Apatite	yellow, pink, blue	pink, yellow, blue
Axinite	light blue, blue, red	light blue, blue, yellow-green
Barite	white	pink, white
Calamine	pale yellow, blue	white, light blue, white-yellow
Calcite	pink, red, orangey-red yellow-green	pink, red
Celestite	yellow, white, bluish-white	white, yellow, pink
Cerussite	yellow, white	yellow, white
Corundum	red, yellow-brown light blue	red, light blue, blue
Diamond	yellow, light blue, green	light blue, green, yellow
Dolomite	red, white, light blue	pink, red, white
Fluorite	blue	blue
Grossular	orangey-red	faintly greenish-yellow, red
Hydrozincite	blueish-white	
Kyanite	reddish	reddish
Magnasite	white, light blue, red	white, light blue, red
Malacon	yellow	yellow
Mimetite	orangey-yellow	yellow
Opal	greenish	
Powellite	yellow	pale light blue
Pyromorphite	yellow	
Scheelite	blueish-white, blue, yellow	pale orange, orange, yellow
Smithsonite	light blue, white, pink	light blue, white, reddish-pink
Sphalerite	yellow	yellow
Spinel	red	red
Spodumene	pink, blue, light blue	red, pink, blue
Strontianite	white, red, blue, pink	white, yellow, green
Thorite		faintly green
Topaz	yellow, green red	light green, pinkish-red
Tremolite	yellowish, orangey-pink, red	yellowish
Willemite	bright green	bright green
Witherite	yellow, white, pink	pink, orangey-green
Wollastonite	pinkish-yellow	pale pinky-yellow
Zircon	yellow, orangey-yellow	yellow, orangey-yellow, pink

3.1.3 OBSERVATION OF MORPHOLOGY, SOILS AND VEGETATION

In an area with a considerable cover of vegetation the morphology of the immediate environment and of the landscape can often give a good indication of the presence of outcrops and of the basic geological structure of the area. Thus lodes, fractures, ridges of quartzite or amphibolite or granitic domes can be marked. The proposed traverse would consequently be modified or a detour made.

If outcrops are rare, the colour and type of soils as well as the type and amount of vegetation, if observed carefully, can give a good indication of the geology and mineralization within the area. A change in the type of vegetation will nearly always indicate a change in the subsurface. One example of this is the 'copper clearings' in Central Africa where the large amount of copper in the soil prevents trees and shrubs from growing there. Furthermore, the existence of certain plants and not others is a good indication of the presence of metal in the soil; thus *Viola utea* var. *calaminaria* often characterizes the existence of zinc. The copper flower (*Asolanthus biformifollus*) accumulates copper in its tissues in quantities reaching 1.3% of its dry weight. It is therefore of interest to the prospector, when a mineral outcrop has been discovered, to notice which plants are growing in the area to help locate other identical mineralizations in the region.

Furthermore, attention should be paid to screes which, when they consist of mineralized rocks or gossan debris, often enable *in situ* mineralization to be found.

3.1.4 LOOKING FOR OLD WORKS

Evidence of old exploratory or productive works is a valuable indication of the existence of a deposit, which can sometimes be exploited further using advanced technology, especially in processing low-grade ore. These works, if they are very old, may not be recognized easily; their identification can be difficult; shafts become filled in, galleries crumble away, ore dumps are partially or totally removed by erosion and the site of the works can be covered or masked by vegetation. However, certain alignments of regular-looking platforms, unnatural looking embankments, piles of ash or slag are indications of the existence of works.

Exploitation sites abandoned relatively recently (less than a century ago) can be recognized fairly easily. The entrances to galleries, almost always masked by bushes, are sometimes betrayed by streams flowing out of them. Dumps are less easy to recognize and if in doubt about the origin of a surface feature, it is best to make minor excavations.

It is helpful, but difficult, to recognize very old works. Old works should not be ventured into alone; someone should be left outside and the length of the exploration decided upon beforehand. Extreme care must be taken during these explorations, which are always dangerous. An excavation should be lit with a naked flame which will extinguish when the air becomes too thin, well before humanly noticeable. The floor of a gallery should be tested with a stick to avoid old shafts with concealed openings and the solidity of the rock face should be checked by hitting it with a hammer which should give a clear ring. The aim of this exploration is essentially to determine the nature of mineralization. This is not always easy because the gallery walls are often covered with a thick layer of heavily encrusted dust particles and the previous miners have often exhausted even the poorest mineral ores.

Samples should be taken at the most favourable places, especially on the floor of the gallery along the lowest level of the mine; for this, however, the floor must be properly cleaned beforehand. A plan of the mine can be quickly mapped out using a compass and pace measurements (or a 'topofil'). Then the old dumps can be examined in order to determine whether they contain minerals which were considered to be of no value at the time of exploitation (tungsten in tin deposits, or blende in lead deposits, mined before the middle of the nineteenth century).

3.1.5 OBSERVATIONS OF OUTCROPS

The following instructions apply to prospecting and not to geological mapping. Observation of outcrops is the most important part of hammer prospecting. Outcrop is continuous only in parts of desert or mountainous country. In most areas outcrops are small, and should be carefully and completely observed as follows:

1) Break the rock in several places, preferably using a heavy hammer.
2) Identify the rock. Remember that structure and texture are most clearly seen on a wet surface. When the determination is doubtful describe the rock and take a sample the size of a packet of cigarettes for a more precise determination and a sliver for a thin section study if possible. Care should be taken when numbering rock samples.
3) Measure the strike and dip of the bedding, schistosity and important fractures, and the direction and plunge of lineations.
4) Look for mineralization with the naked eye and where necessary use a hand lens; see if the rock has been altered; look for products of oxidation. Mineral deposits do not usually appear at the surface, especially in the tropics. Even in the most favourable cases their outcrops consist only of a few square metres of gangue, commonly sterile, as most minerals of metalliferous deposits are sulphides that are readily oxidized, and whose alteration products are more or less soluble in rainwater. Oxide minerals are more resistant. They include magnetite, haematite, chromite and some

Table 2.3. *Colour of minerals at outcrop.*

Mineral or metal	Outcrop colours	Oxydized compounds
Iron sulphides	yellows, browns, chestnuts, reds	Goethite, hematite, limonite, sulphates
Manganese	blacks	Mn. oxides wad
Copper	greens, blues	Carbonates, silicates, sulphates, oxides, native Cu
Cobalt	Black, pink, sometimes violet	Oxides, erythrine
Nickel	greens	Annabergite, garnierite
Molybdenum	bright yellow	Molybdenum oxides, iron molybdate
Silver	waxen greenish	Chlorides etc. Native Ag
Arsenic	greenish, greens, yellowish	Iron arsenate
Bismuth	yellows	Bismuth-ochres
Cadmin (in zinc)	light yellow	Cadmium sulphide

manganese minerals, notably braunite. While certain brightly coloured salts do exist which, by colouring the gangue, indicate the presence of some of the metals, they are seldom abundant. **Table 2.3** taken from Bateman (1951) lists some of the colours given by certain metal outcrops. **Table 2.4** gives the colours of the streaks of mineral ores. Attention should be focussed on altered and decayed rocks, and rocks coloured red, brown or black by oxides of iron or manganese. If they are composed partly of cavernous quartz, more or less impregnated with ochreous clay and haematite, there is a good chance that they form the ferruginous capping (gossan or 'iron hat') of a mineralized body. Closer study will show whether or not minerals or only their imprint (boxworks) are present. Certain petrographic changes in rocks are directly related to the emplacement of minerals, having occurred at the same time, and for the same reasons, as the mineralization. They therefore act as a guide to the presence of minerals. 'Tourmalinazation', the impregnation of rocks by tourmaline during contact metamorphism, is common near tin deposits. 'Greisinization' is the conversion of granitic rock into an aggregate of dominant quartz with lithium-mica, and also indicates the presence of tin. Kaolinization the conversion of feldspar into white clay, is specific to felsic igneous rocks (granite, trachyte), and occurs particularly in association with tin-bearing zones but is also common in association with veins mineralized with gold

Table 2.4. *Colour of minerals' streaks*

White	*Black*	*Appearance*
Silver	Magnetite	Arsenopyrite
Bismuth	Pyrulosite	Pyrrhotite
Calamine	Stephanite (metallic black)	Stibnite
Pyromorphite	Tenorite	
Sphalerite sometimes		*Grey*
	Green	Cerargyrite (shiny grey)
Blue	Malachite	Galena (lead grey)
Azurite		Molybdenite
Malachite	*Brown*	Platina
	Cassiterite (brownish)	Tetrahedrite (steel grey)
Yellow	Chromite	
Calomel	Franklinite	*Red*
Limonite	Hematite (brown yellow)	Cinnabar
Gold	Iron pyrites (blackish brown)	Cuprite (red brown)
	Sphalarite (reddish brown)	Native copper
Greenish black		Niccolite (pale red)
Chalcopyrite	*Blackish grey*	Pyrargyrite
	Arsenopyrite	Siderite (dark cherry red)
Blackish	Chalcocite (blackist lead grey)	
Asbolite	Chloanthite (metallic appearance)	
Argenite (metallic appearance)		
Stannite		

or gold and silver. Dolomitization of calcareous rocks is in some cases an indicator of the presence of calamine, haematite and manganese oxide in limestone.

5) Should significant mineralization be found, give an estimate of the percentages of the minerals of interest (**Fig. 2.6**), note the form of the ore body and, if present, the gangue; eg vein along a fracture or bedding vein, along a fault and in contact with rocks of different kinds, stockwork, impregnation by the ore mineral with or without gangue. Note the nature of the contact of a vein with the country rock, is a selvage present? The next step is sampling.

3.1.6 STRIPPING OUTCROPS

In addition to the work which often has to be carried out on the surface in order to study a mineralized outcrop, the prospector must note the characteristics of the discovered indication by a series of excavations especially in the case of an isolated occurrence in a region with limited accessibility. These operations consist essentially of three strippings or trenches:

1) The first trench cut into the indication itself is aimed at finding out the geometrical characteristics of the outcrop: dimensions, thickness in particular if it concerns a lode, strike, dip. This process is made easier if a bulldozer is used. This sort of equipment is being used more and more for surveying indications (see below).
2) The two other surface cuts about 10m from each other and from the first cut on the assumed continuations of the occurrence and perpendicular to its strike; they should as far as possible reach bedrock. They are intended to show that the deposit continues longitudinally and to see if the occurrence found, often poor due to leaching of the outcrop, is rich throughout.

We are only concerned here with very light operations intended essentially to provide a few necessary pieces of information in order to fill in the index card for the occurrence. Prospecting by systematic surface cuts will be dealt with later on.

These operations should not last more than a week, even if they have not produced evidence of further data. The results obtained, even if they do not indicate mineralization, must be noted on the occurrence index card.

3.1.7 SAMPLING

At this stage of mining research sampling is qualitative rather than quantitative. Various cases can arise:

1) If the mineral outcrops, take several samples so that polished sections can be studied, proceeding as with the petrographic samples, and several samples each of 100g to 2kg for analysis. The more heterogeneous the material, the larger the sample taken. All samples must be as representative as possible of the degree of mineralization.
2) If there is a gossan take 10 to 15 samples throughout the gossan and from

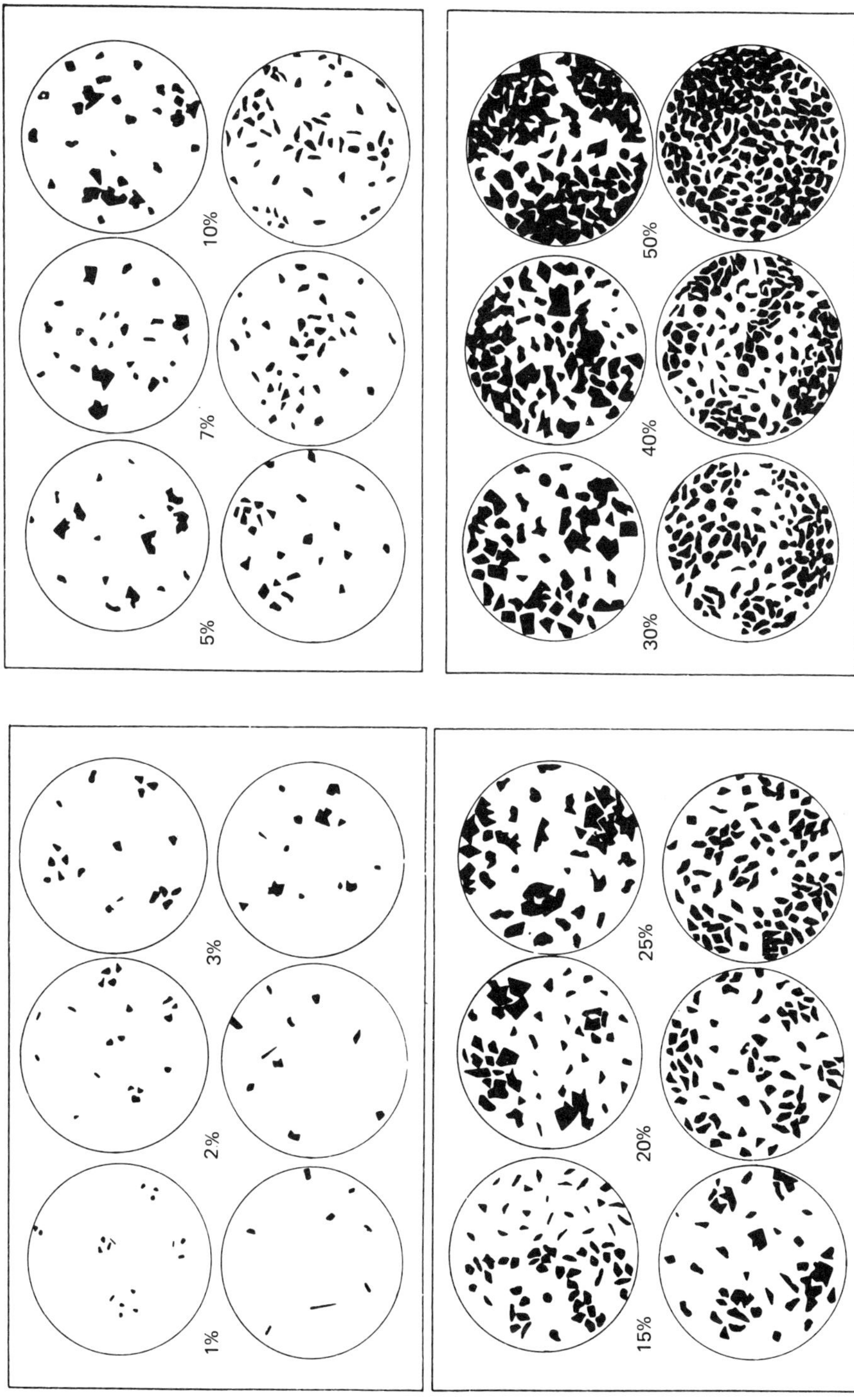

Fig. 2.6. *Relative percentages of mineral grains under the microscope (from Terry and Chilingar, 1955, Journal of Sedimentology–Petrology* **25**).

as many different facies as possible; ask for a multi-element analysis since the presence of trace elements often helps establish whether or not gossan is derived from useful sulphides.

3) In the case of pyritic rock with disseminated sulphides, a rock which has undergone a hydrothermal alteration and a rock containing disseminated oxidized products, it is recommended that, using a hammer, a chip sample of rock is taken, avoiding areas where minerals are visible. It is obvious that the objective is only qualitative and that the grades which are supplied by the analysis should not be taken literally. Collect one or more samples of 2 to 3kg on average, taking rock from numerous sites over the entire mineralized outcrop.

Place the samples in canvas or plastic bags. The sample number should be written immediately on the bag and also on the plastic label slipped inside the bag. A record must always be kept so that any losses which may occur during transportation can be replaced.

For the preparation of the sample you are advised to make an arrangement with the laboratory in charge of the analysis. You have two choices. The first is to grind the sample in a cast-iron mortar for a short time, quarter it in the sample-divider and send 500g to the laboratory which will prepare the powder. The second is to quarter it until you have a 50g sample. Grind it completely to eliminate lumps, sieve it (through a mesh) depending on the type of analysis required (presently 40μg for quantometric analysis and 80μg for atomic absorption and mass spectrometry) and send the powder to the laboratory.

Packaging the sample and the analysis request are described in Chapter 5. Preferably ask for a quantitative analysis or a mass spectrometry analysis. Then, for those elements which reach the upper limit of the spectrum, it is always possible to ask for a more precise analysis using atomic absorption for example.

3.1.7.1 Loaming and dollying (soil and quartz panning)

In the case of quartz veins with invisible gold-bearing mineralization, a mineralogical examination of the quartz is necessary. If the quartz is compact without traces of sulphides or their oxidized products or even cavities covered with iron coatings, there is no need to delay over the occurrence. If examination of the quartz has shown the presence of sulphides or their residues, carry out soil panning and possibly quartz panning.

Soil panning: This operation is pointless if visible gold has been observed. Take some soil and quartz debris from a depth of 5 to 30cm of the vein outcrop or from its wall or from between two blocks of quartz of an alignment (for the panning technique, see Chapter 3). Wash the contents of the pan and note the number of spots or the presence of colours (strong or weak). The operation is repeated at several points along the outcrop or alignment of blocks. The result is considered positive if colours have been constantly observed for a certain number of points sampled. On the other hand, colour observation in a single pan has hardly any significance.

Quartz panning: This operation is carried out if the soil panning result was positive, if the vein showed visible gold or on screes where soil panning is impossible.

Take a sample of 500g to 1kg of quartz, weigh the sample, then grind it in an iron mortar to 0.160mm (80 to 100 Tyler mesh). The ground sample is panned; the gold is separated (possibly by adding a drop of mercury) and weighed. The operation is repeated at a certain number of points, however, you do not need to worry about the samples being representative. The soil and quartz panning technique is above all used for gold prospecting but it can also be employed for dense mineral exploration (cassiterite, wolfram, etc.) finely disseminated in various environments and not just in veins.

3.1.8 KEEPING A FIELD NOTEBOOK

All observations are noted in the field notebook. If you do not have a loose-leaf book, it is a good idea to reserve the left hand pages for the sample and sketch numbers. Obviously the observations made of the outcrops will take up most space: general description of the outcrop, measurements, description of the rocks, main fractures and mineralization, note of the sampling method and sample numbers. If an occurrence is discovered, the description should be accompanied by a sketch and a cross-sectional drawing.

3.1.9 UPDATING DOCUMENTS

Different maps are drawn up:

1) Traverse map where observation numbers refer to the field notebook.
2) Outcrop map indicating with symbols the nature of the rocks, strike and dip of the deposit and the schistosity; this map will always be a useful document, both in order to familiarize yourself with the environment of the occurrences and to be subsequently taken up again at the time of routine geological survey; a special symbol denotes float, if the region is poor in outcrops.
3) Sample map which could include a table of analysed results.
4) Mineralization map showing all the previously and recently discovered mineralization.
5) Discovery of an occurrence will require that index cards are completed (Appendix 1).

3.1.10 PROSPECTING BY HELICOPTER

A helicopter is an excellent means of strategic prospecting in regions with little or no vegetation cover and broad zones or outcrop. It enables the observation of large areas in a relatively short time and the outstanding features show up much better and with more continuity than from a low angle of view. Their colour and form are accentuated and this is increased if the correct altitude and angle of sun are chosen. These depend on local conditions and are determined by several tests at the beginning of the expedition. For example, in a hot arid region full sunshine (between 10.00 a.m. and 2.00 p.m.) is unfavourable.

Observation from a helicopter allows hydrothermal alterations, gossans, old workings, specific horizons (chert, jasper, carbonates, conglomerates, etc.), veins, fault zones, contact zones, mineral lenses (chromite) and mineral beds to be easily identified. The environments and the occurrences can therefore be very varied and in view of the cost of the helicopter, it would be advisable to take a fairly comprehensive set of prospecting equipment if space permits (scintillometer, magnetometer, etc.).

In practice, when prospecting it is preferable to have two others in the helicopter besides the pilot, one paying attention to the location while the other records the observations. In fact, the work is organized according to circumstances, taking into account the local conditions and the type and grid of prospecting. The helicopter allows work to be carried out from point to point but also following more or less long sections, setting down at one place and picking up at another. If it has sufficient capacity it could allow several groups to work at the same time. Due to its high cost work should be carefully planned in advance and detailed photographs taken each day.

3.2 Studying occurrences and surveying occurrences on anomalies

Studying occurrences aims to judge the significance of a discovery with a view to proceeding to more detailed prospecting: tactical, geochemical, geophysical, digging trenches or shafts and sometimes even direct drilling. Surveying occurrences on anomalies consists of studying the mineral *in situ* from geochemical and geophysical anomalies or alluvium. At this stage of the survey you generally know which substances are best surveyed in the selected zone. More often than not you need to determine the extension of a mineralization from one or more isolated occurrences. The value of this mineralization will be judged on its known or supposed size and the mineral grade. The work focusses principally on two points:

1) Drawing up a geological map if there are enough outcrops or an outcrop map to serve as a framework and guide for prospecting.
2) Carrying out careful sampling to be as exhaustive as possible.

3.2.1 BUILDING UP A GRID

In some cases you can still make use of the grid from geophysical or geochemical prospecting or an enlargement of an aerial photograph or a cadastral survey. If not, first make a systematic tracking from a base line. For example, for a 1:2000 survey the tracks are spaced 40m apart and a marker is placed at 20m intervals along each track. Compass sightings are taken with possible slope corrections using a clinometer. Sometimes it is preferable to make a topographical map using a clinometer ruler.

3.2.2 CARRYING OUT THE SURVEY

This is no different to the survey of prospecting occurrences: observation of float, outcrops, relative measurements of stratification, schistosity and fracture, determination of mineralization (percentage). In a region where the density of outcrops is low it is necessary to cover all the water courses and gullies in the search for outcrops. On a sufficiently advanced and promising subject where the formations are strongly deformed a structural analysis will often prove to be essential. At this stage of prospecting you can often benefit from guidelines which should be followed very carefully: distinctive structure (fold hinge, fractures), distinctive horizon (carbonate, graphite, jasper or chert, conglomerate), distinctive palaeogeography (channel, transgression), position of the mineralization in relation to an intrusion, lode, fault zone, hydrothermal alterations, vegetation, etc.

The format of the field notebook will be exactly the same as that for occurrence prospecting but the sketches and sections are more important. An occurrence index card should be completed (Appendix 1).

3.2.3 TRENCHING

Trenches are dug especially when the width of the mineralized bodies is the smallest dimension, and when they are covered by several metres of earth and alluvium. Digging out can be done manually (using spades, picks, possibly pneumatic drills or explosives) but there is an increasing tendency to use automatic diggers. These machines are often hired out to companies which use them in the areas where the surveying of occurrences is carried out. It is advisable to hire the most adaptable machines, in other words not too big. The advantage of a bulldozer is that it can be very widely used but it has the disadvantage of having a very wide blade forcing it to clear huge volumes of earth for a single trench. For this reason it is best wherever possible to use hydraulic shovels allowing trenches to be dug out (such as Poclain retro hydraulic shovels or JCB); the models which have buckets with a maximum capacity of 500 litres are best suited to surveying occurrences.

The use of one method in preference to another will depend essentially on the local conditions (situation, land use, ease of hiring and moving machinery, etc.). As a rule, the trench should be dug perpendicular to the probable alignment of the occurrences or to the long axes of the geochemical anomalies.

When the mineralization is marked by a scree of minerals on a slope the trench is made perpendicular to the alignment of the screes at the highest point up the slope. It should begin a few metres below the upper screes and continue upwards from them for varying distances. They should stop when one of the following conditions has been fulfilled:

1) when the top of the slope has been reached;
2) when the trench passes into ground different from the rock of the mineralized scree, if they are composed mainly of mineralized rocks and not of gangues;
3) and above all when mineralization outcrops.

If none of these conditions is fulfilled, the trenches will be stopped after 50 to 100m to be taken up again if nearby work shows a need for it. It should be noted that in a humid tropical environment there is a very real danger of landslides. Precautions must be taken to prevent any accidents. In certain cases, for example a quartzose outcrop in laterite without a definite strike, you must begin by stripping the outcrop until you discover the trend of the walls. Two stakes planted in the uncovered axis of the lode at the edge of the stripped zone will give the general trend which will be plotted out on the ground by stakes aligned by the two preceding ones. Trenches will be dug perpendicular to this trend and centred on the stakes at a distance of 20 to 50m from the original outcrop. They are intended to confirm the continuity of the lode and each one should not exceed a length of 10m.

In areas with dense cover, or when there is insufficient data to determine the direction in which to dig the trenches, you must begin by digging two trenches which intersect at right angles. The length of the trenches should be fixed according to the local conditions. A width of 1m is sufficient; as for the depth, it should be deep enough to reach the rock *in situ*. During the digging of trenches the earth with visible mineralization should be carefully separated from the sterile earth by shovelling one lot to one side of the trench and one lot to the other. These operations and the information they provide will serve to complete the existing index card or to start a new card if it is a new occurrence.

3.2.3.1 Observations

As with outcrops the information obtained from the trenches is concerned with the mineral and its gangue, the country rock and the relationship between the following elements:

1) Mineralization and gangue: note the nature of the different minerals and the gangue and also the way in which the valuable minerals are dispersed within the gangue: longitudinal veinlets, banded or unbanded, transverse veinlets, anastomosing veinlets, patches or crystals, bunches, superficial coatings. The thickness of the mineral veinlets or the visible surface of the patches, crystals or bunches should, as far as possible, be noted as well as, in the case of crystals, patches or bunches, their number per unit of surface area of the outcrop (square decimetre or square metre according to the extent of the outcrop).
2) Country rock: note the nature of the country rock, if it is recognizable, its degree of alteration and the possible presence of distinctive minerals or mineralization. In most cases this rock will have been transformed into superficial products of alteration and the nature, colour, etc. of these should be noted. If outcrops of rock can be seen a short distance away from the occurrence they should be noted indicating their nature, strike and dip of the deposit if they are sedimentary rocks, or of lineations if they are metamorphic rocks or tectonized eruptives, and the strike and dip of the major joint sets. Beforehand you must ensure that the outcrop is really *in situ* and do not forget to note down whether it is sub-*in situ* or *in situ*. In the first instance, trends and dips can be given with qualifications, particularly confirmation from outcrops not far away that show similar characteristics.
3) Relation between the mineral and country rock: this is essentially a question of noting the way the mineral has been deposited and its possible gangue in the country rock.

3.2.4 SAMPLING

If the trench intersects a mineralization or the signs of a mineralization (gossan, hydrothermal alteration, disseminated pyrite), samples will be taken by channel sampling or cut with a hammer and chisel: vertical channel sampling if the mineralization is linked to a stratigraphic level or to a horizontal bedded vein; horizontal channel sampling at the lower part of the trench, if the mineralization is stratiform and folded, or linked to a vertical vein or is disseminated. All quantitative sampling should be carried out with the greatest care and should be closely supervised.

3.2.4.1 Sampling of a vein

The vein is exposed for a height of about 50cm. Then the decomposed, fractured upper part is removed. After stripping, the sampling proper begins (**Fig. 2.7**). The operations are carried out in the following manner. The surface to be channelled is

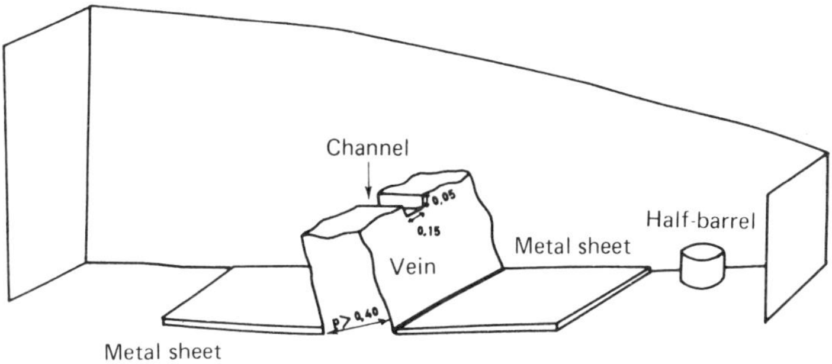

Fig. 2.7. *Sampling a vein.*

first made more or less smooth (perpendicular to the walls), then carefully scrubbed, using water and a brush. Two parallel lines are then drawn 15cm apart with white or coloured industrial chalk on the surface perpendicular to the walls. The cut is made with a chisel, removing 2kg over a uniform depth of 5cm.

It is important not to cut too deeply into the soft rock or too little into the hard rock; the differences in hardness generally correspond to the differences in mineralization. Irregularities in the channels may therefore give a false idea of the grade; most often, the mineralized parts being the softest, the apparent grades obtained are higher than the real grades. To make sure that the channel is at a constant depth it is helpful to use a wooden template (**Fig. 2.8**).

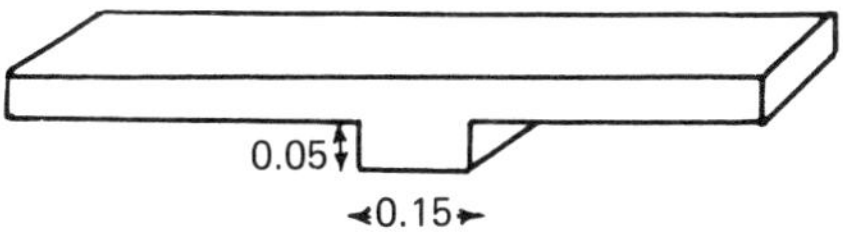

Fig. 2.8. *Wooden template.*

The fragments of vein rock broken off by the chisel are carefully gathered on to metal sheets placed on the ground on either side of the vein. Care must be taken to gather all the fragments. In the absence of metal canvas sheets may be used instead. It is essential that the samples are never laid down on the ground itself. When the channel is finished all the fragments gathered on to the metal sheets are collected and placed in a receptacle (half-barrel, pan, small wash pan, basin, etc.).

The metal sheets should be carefully brushed down and the dust put into the same receptacle. The receptacles containing the samples are duly numbered, removed from the trench and the next step is the reduction of the samples to a volume judged to be acceptable for transportation. The intervals between channels will vary from case to case; they may vary from 1 to 5m. In the case of bunched mineralization (eg wolfram) the channels may be spaced 5m from one another but they will be bigger (width 50cm and depth 10cm). Possibly, to cut down the work involved in sampling a blow-torch may be used to heat and break up the rock.

3.2.4.2 In the case of a more or less smooth outcrop

The technique of channel sampling on an outcrop which is more or less smooth does not differ from the one used for veins. The channels are cut perpendicular to the alignment of the mineralization. Each sample represents a length of channel varying from 1 to 5m (Chapter 7).

3.2.4.3 Sampling by recovery of bore hole cuttings

When ground conditions permit, ie when it is not too hilly and vegetation is sparse, as particularly in the African Sahel, access to any site may be gained without difficulty using a cross-country vehicle towing a compressor.

With a hammer drill and the corresponding dust trap, particularly the Atlas Copco DCT 31, a section 1 to 1.5m thick may be sampled in an outcrop without stripping, if the aim of the operation is merely to find the levels of grade. For a more detailed sampling a trench must first be dug. When dealing with an extended vein outcrop a section is made perpendicular to the outcrop at intervals of 10 to 20m depending on the length of the vein. Along each section holes are drilled on average every 1m or 1.5m across the width of the outcrop; when no stripping is done, it is possible that sampled with the *in situ* outcrop, are eluvial boulders surrounding it but this is not important in view of the aim of the operation. Holes are bored at an inclination of about 45° to the horizontal, the opening of each hole being located approximately vertically above the previous hole according to the previous holes inclination and width. At the opening of the hole an air vent, through which the drill bit passes, collects the cuttings which are sucked up by the dust trap through a flexible hose. The cuttings from each hole are collected in a bag attached to the base of the trap. Once the hole is finished the blades of the trap are shaken and the bag of cuttings is removed and replaced by a new bag. The inside of the trap is carefully cleaned only when all the holes belonging to the section have been completed. Each bag containing the bore-hole cuttings is sent to the laboratory for analysis. It carries the usual information of the site, section number and bore-hole number which is also the sample number.

3.2.4.4 In the case of a vertical wall

In the case of a wall (**Fig. 2.9**) showing slightly inclined or horizontal mineralized beds sampling can be carried out using vertical channels at intervals varying from 1 to 5m. These channels must be 15cm wide and 5cm deep (Chapter 7). One or more samples can be taken along the length of each channel. Only one sample will be taken if the mineralization appears to be of uniform distribution in only one rock facies. If several samples are taken, either following visible changes in mineralization, or because of variations in the surrounding rock facies, the limits of the various

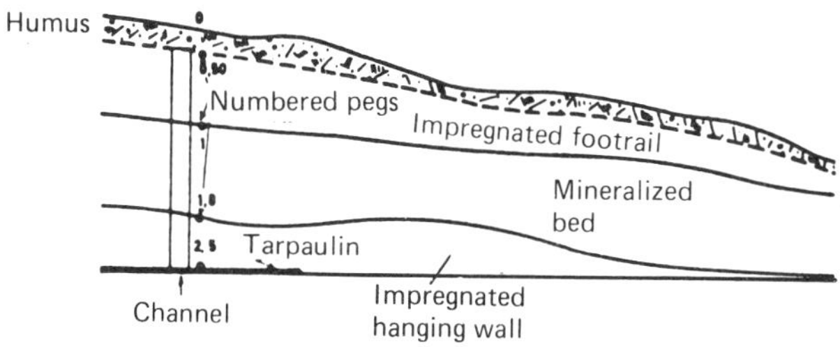

Fig. 2.9. *Sampling an ore deposit.*

samplings will be made clear by the embedding of a numbered peg beside the channel, the number corresponding to the depth reached by the wall at the point in question, measured in centimetres from the surface.

The substances removed from the channel can be gathered up either directly into a receptacle or bag held by a worker, or on to a metal sheet or a tarpaulin laid out on the surface of the trench. This metal sheet or tarpaulin must be carefully cleaned after all of the fragments and dust obtained from each channel or channel section have been emptied into a half-barrel or bag. A label with the sample number must be placed in the bag or half-barrel.

3.2.5 SAMPLE PREPARATION

3.2.5.1 Crushing

Samples taken during hammer prospecting generally have a volume of less than 20 litres. It is therefore sufficient to crush them using a pestle in a cast iron mortar after having broken up the largest blocks with a hammer. This will reduce the grain size to less than 1mm.

3.2.5.2 Reduction

In the majority of cases the sample taken is too bulky to be sent to the laboratory. It is necessary to reduce it, but to do so in such a way that the reduced sample obtained is identical (or at least not very different) to the original sample taken. Reduction is a very delicate operation, and should always be carried out under the constant supervision of the prospector. The simplest sampler is a cross of thick metal sheet (3 to 5mm) for dividing the heap to be sampled into four equal parts. Usually the Jones sample-divider is used: a metal box divided off by transverse gutters pointed alternately from one side to the other of the box. Three models of sampler of different sizes exist, depending on the grain size of the sample, but No 6 is sufficient for a small load of fine ore up to 4mm.

The sampler's accessories consist of two tubs, which are placed beneath the gutters on either side, and a feeding shovel the size of which is adapted to that of the sampler. All the accessories are made of metal, which makes cleaning and brushing easier. Jones sample-dividers are specially designed for powder and dry particles. For a rapid and accurate operation, in addition to the sample-divider, either three boxes specially adapted for it, or normally two boxes and a feeder shovel must be available; two of the boxes being in place on either side of the device, at the foot of each series of gutters, the contents of the third box are emptied into the sample-divider **(Fig. 2.10)**, taking care that the flow of ore is even along the edge of the box. If the batch to be sampled is of a greater volume than the capacity of the box, then the box is filled up and emptied into the sample-divider as many times as necessary to use up the whole batch, the volume of which should in any case be less than that of the two empty boxes. The box is carefully brushed out above the sample-divider, then set down on the ground. The sample-divider is then shaken slightly and brushed out to remove all the particles.

Next, one of the full boxes below the gutters is taken away and replaced by the empty box. If the boxes are large enough to hold all of the original batch of ore, the other (third) box is not emptied, except in the last two quarterings if one or several

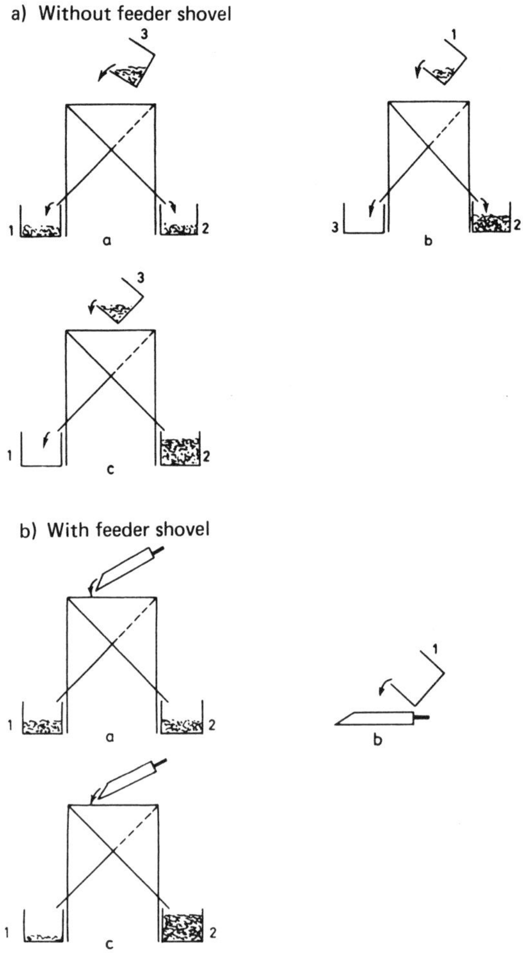

Fig. 2.10. *The use of the Jones sample-divider.*

reference samples are to be kept or panning is required. In that case, the different quarterings are carried out always using the boxes from the same side of the sample-divider. If the capacity of the boxes is not sufficient, then there will be a case for emptying one of the two receptacle boxes on the ground away from the local vicinity, either after each quartering or after two or more quarterings depending on its capacity. Each time it is emptied, the box will have to be brushed out in case it is contaminated with ore from the quartering.

If in possession of the feeder shovel belonging to the normal sampler equipment, after the first quartering carried out on the original sample poured on to the shovel, the shovel needs to be thorougly cleaned over the device and the sample-divider itself. The contents of box 1 are emptied on to the shovel and tub 1 is put back in place. The contents of the shovel are emptied into the sample-divider and the process starts all over again. Care must be taken in the transferals from the tub to the shovel not to let any ore fall off the shovel.

The sample obtained will weigh 100 to 200g. It either can be sent directly to the analysis laboratory which will take care of its final preparation, depending on the kind of analysis required, or, if the transport costs are high, it can be prepared on-the-spot to obtain the final powder.

3.2.5.3 Grinding and screening

In this last case quartering will be extended until a sample weighing 50g is obtained, which will be ground and screened allowing no screen oversize. The grain size to be obtained depends on the type of analysis required (at the present time 40μm for quantimetric analysis and 80μm for wet chemistry, atomic absorption and emission spectrometry).

In hand-grinding precautions must be taken especially concerning grains or dust falling from the mortar, having been carried along by the upward movement of the pestle. There are two remedies:

1) do not overfill the mortar;
2) move the pestle less and do not raise the head of the pestle over the mouth of the mortar.

On the other hand, in every ore there are particles which are harder than others. It will therefore be necessary to screen quite frequently to end up with sufficiently ground particles in the mortar, and do not forget to recycle the screen oversize. After grinding samples care must be taken to clean the pestle and mortar thoroughly, brushing the dust on to the heap of ground material.

3.2.6 SHAFTS

Here we describe small exploration shafts or shafts for sampling horizontal or subhorizontal tabular deposits or occurrences (bauxites, phosphates, etc.). They may be isolated or disposed following a square or rectangular grid network, depending on the stage of prospecting.

Usually these shafts are sunk by hand and their opening is circular with a diameter of 0.70 to 0.80m. The supervision of the sinking of the shafts and the precautions to be taken are the same as for shafts in alluvium (Chapter 3).

Sampling in these shafts may be carried out using channels or by using the rock extracted per metre of depth. Procedures for channel sampling are identical to those used in trenches, in the case of a vertical wall (**Fig. 2.9**).

The procedure described on the rock extracted is used for ores of homogeneous composition and low market value (bauxites, iron and manganese ores).

Soil overburden is removed beforehand from the shaft site and an area of radius 5 to 6m around the axis of the shaft is cleaned.

The measured heaps are laid out in the form of a spiral around the shaft clockwise (**Fig. 2.11**) and separated from one another by logs or wooden planks on the site, the first heap being placed near the shaft's numbered peg. Throughout this book the measured heaps can be taken to be 1m high.

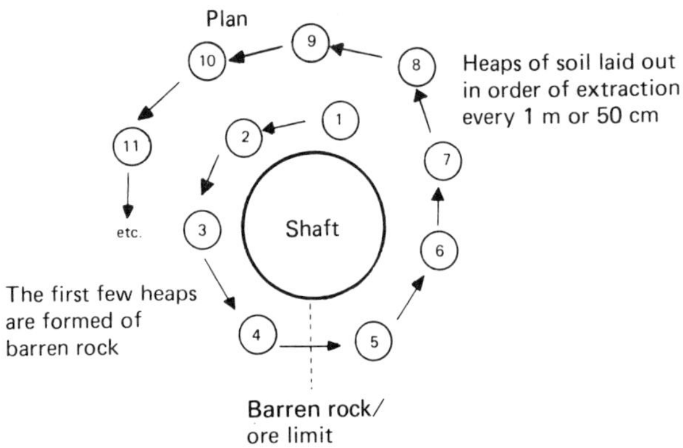

Fig. 2.11. *The arrangement of measured heaps (im high) around a prospecting shaft.*

It is generally useless to arrange a plank on the ground beforehand. Small pegs are planted on every heap with a little board carrying the following information: shaft designation, metres of depth corresponding to the heap.

Reduction of a sample with a small volume is carried out as before, except that for samples which start off being bulky crushing, and if necessary, grinding are generally carried out using mechanical devices. Of the petrol-driven laboratory crushers the best known is the Knapp and the Bates which exists in several sizes. Size 1 is for loads of 45 to 120kg/h, size 2 for 120 to 220kg/h and size 3 for 220 to 500kg/h. The crushed ore is recovered in a box which may be one of the Jones sample-divider metal boxes.

The complete batch should be crushed before going on to the sample-divider. After crushing the gutters and the jaws of the machine should be brushed out and the dust collected in the box containing the ore before it passes through the sample-divider. Laboratory disc grinders are rarely used in the field.

The crushing and grinding meshes to be used for each reduction of sample weight by quartering may be determined approximately using the sampling method in Fig. 7.21 (Chapter 7).

3.2.6.1 Motorized auger sampling

For showings or deposits of certain plastic or soft substances, such as talc, light motorized augers are used which can take samples to a depth of 15m by adding a small 5 to 6m hoist.

3.2.7 SAMPLING NOTEBOOK

The final sample is sent to an analysis laboratory complete with its label which contains all the necessary information. This label remains with the sample throughout all preparation stages. Each prospector keeps a sampling note book in which he notes all the samples which he has taken, whether they are analysed further or not, in numerical order and using a maximum of four numbers. The sample number can be made up, for example, of the prospector's initials, followed by the year the sample was taken (only the last two numbers) and finally, separated by a dash, the number of the sample itself. However, there are other ways of numbering samples and these are often specified for each expedition or territory.

Each sample can be denoted twice by placing carbon paper between two successive pages, the original is then detached and used as a label for the sample and the copy remains in the sample notebook. The following information should be noted in the sample notebook for all samples:

1) sample number;
2) name of area or locality;
3) the actual location of the sample, as precise as possible; it is not sufficient just to give the coordinates, they should be given in relation to striking topographical features;
4) if possible the nature of the sample; it is generally better to name the sample even if this means following it by a question mark when in doubt.

For samples intended for analysis which give rise to quatering the following information should be added:

1) sampling method (grab and channel sampling);
2) weight of the crude sample, if it is not possible to weigh the sample its weight is estimated and 'estimate' is written after it.

When sending the sample to an analysis laboratory it is preferable to put it in a lined bag made of plastic or canvas. The sample number is marked on the bag and also on a label slipped inside the bag.

References

Bateman, A M (1951) Economic mineral deposits. John Wiley and Sons.
Coldmer, F (1920) Prospection, recherche et etude des gites miniers.
Goodwin, W L (1956) A handbook of prospecting National-Business Publ. Gardenvale. CY, P Normogramme d'echantillonnage. Minerais et metaux.
Kuzvart, M; Bohmer, M (1978) Prospecting and exploration of mineral deposits. Developments in Economic Geology, 8, Elsevier Publishing Company, references on deposits and mining exploration for most of Eastern Europe.

McKinstry, H E (1948) Mining geology. Prentice Hall.
Peters, C W (1978) Exploration and mining geology. Department of Mining and Geological Engineering. University of Arizona, references on deposits and mining exploration world wide.
Savage, E M (1934) Prospecting for gold and silver. McGraw Hill.
Von Bernewitz, M W (1934) Handbook for prospectors and operators of small mines. McGraw Hill.

Appendix 1

Sample file sheet

Date established:

by:

Number:

Name:

Principal elements:

Structure (vein, stockwork, mass, stratum, shear-zone):

Location and access:

Latitude: Longitude: Altitude:

Geographical and geological maps:

Local geology (surrounding rocks) and regional geology from petrographic, stratigraphic and structural point of view:

Description of occurrence:

Length: Width: True thickness:

Direction: Dip:

Structure of ore (massive, disseminated, bedded):

Mineralization: gangue:
primary minerals:
supergene minerals:

Type and relationship with surrounding rocks:
(sedimentary, volcano-sedimentary, acid differentiate, segregation, skarn)

Alteration in surrounding rocks:

Results of analysis:

Previous work (brief description and sketch):

Work carried out (description and sketch):

Other observations:

Bibliography:

Chapter 3

Prospecting alluvial, eluvial and coastal deposits: Methods and economic evaluation

Alluvial prospecting, although sometimes considered an archaic method, is very efficient, if gravels are sampled very carefully by a person experienced in general alluviation processes. This is undoubtedly a prospecting method where experience determines the results obtained. This method of prospecting consists of taking and washing alluvial and eluvial samples to extract useful minerals and evaluating their concentration levels (expressed in g/m^3).

1 Strategic alluvial prospecting (regional reconnaissance)

The goal of strategic prospecting, whether alluvial or geochemical, is to locate points (showings or anomalies) worth examining in a later phase. As a rule, geochemical prospecting and alluvial prospecting are carried out simultaneously in the drainage course. Two types of zone are of interest: zones containing anomalies and mineralization where tactical, essentially geochemical, prospecting will be performed (Chapter 4); points of the drainage course with marked alluvial mineralization of certain substances (Au, Sn, Nb, Ta, W, diamonds, gemstones), these points of interest are checked by alluvial prospecting which is traditionally called general prospecting. This chapter only describes the methods for checking and analysing points of interest in general prospecting.

1.1 General requirements

The sampling density must be as uniform as possible, since most of the samples are located in the tributaries and subtributaries of the main drainage course (the latter could be ignored at this stage of exploration). They are always taken upstream from the confluence, unless lack of time means that the number of samples is reduced, in which case only one collective sample is taken downstream.

Sampling must be carried out at the optimal concentration points for heavy minerals: rock sills, pot-holes, large pebble deposits, river bed narrows, etc. At the selected sampling point, especially in the case of torrential type deposits on uneven bedrock, it is best to take two or three samples at intervals of 1 to 2m or 12m, for subsequent mixing to provide an average sample of the deposit.

The volumes of alluvial material sampled must always be measured and numbered: the volume and number are recorded on the label which accompanies the corresponding pan concentrate. A sampling index card is prepared, and completed subsequently with figures showing the economic mineral content or 'tracer' minerals that are useful in completing the geological map.

1.2 Plotting traverses

1.2.1 PURPOSE

To cover a maximum area efficiently, thereby reducing time and costs to a minimum.

1.2.2 MEANS

Systematic, so-called strategic, prospecting of streams by taking alluvial and geochemical samples in the active stream bed (see Section 1.2.3).

1.2.3 METHODS

Plot traverses every 20 days maximum or about every 160km, since experience has proven that average progress in a forest is 8 linear km per day. Plot these traverses to a scale of 1:200,000, bearing in mind that **(Fig. 3.1)** each river basin must be 'covered', at least by the main drainage course and the confluence with its tributaries; no 'ridge' must be left 'uncovered', ie one whose streams have not been sampled; the time allotted and the number of teams available are accounted for; (this factor sometimes entails looser traverses with blank areas); the photogeological map must be analysed and, if necessary, traverses planned on outstanding points (contacts, showings reported, etc.); when the general traverse is well-plotted to a scale of 1:200,000 it can be transferred to 1:50,000 scale maps for use in the field.

1.2.3.1 Plotting to 1:50,000

1) Calculate the exact distance using a map measurer. Locate the alluvial sampling point according to the grid requested by the expedition order for the general survey programme (in strategic prospecting the alluvial sampling is always duplicated by a geochemical soil sampling for example, in the equatorial forest, for sedimentary formations, one alluvial sampling is made every 2km and one geochemical sampling is made every 1km; on the ancient basement zones one alluvial sampling is taken every 3km and one geochemical sampling is taken every 1.5km). Follow the above grid for the main drainage courses. At each confluence, locate one alluvial sampling point for each tributary longer than 2 to 3km (depending on the grid selected) and always locate a geochemical sampling point 1km upstream.
2) Position one geochemical sampling point on the tributary shorter than 2 to 3km (depending on the grid selected). Count the number of alluvial and geochemical sampling points per traverse, where each alluvial sampling point also has a geochemical sampling point (bed + right lateral + left

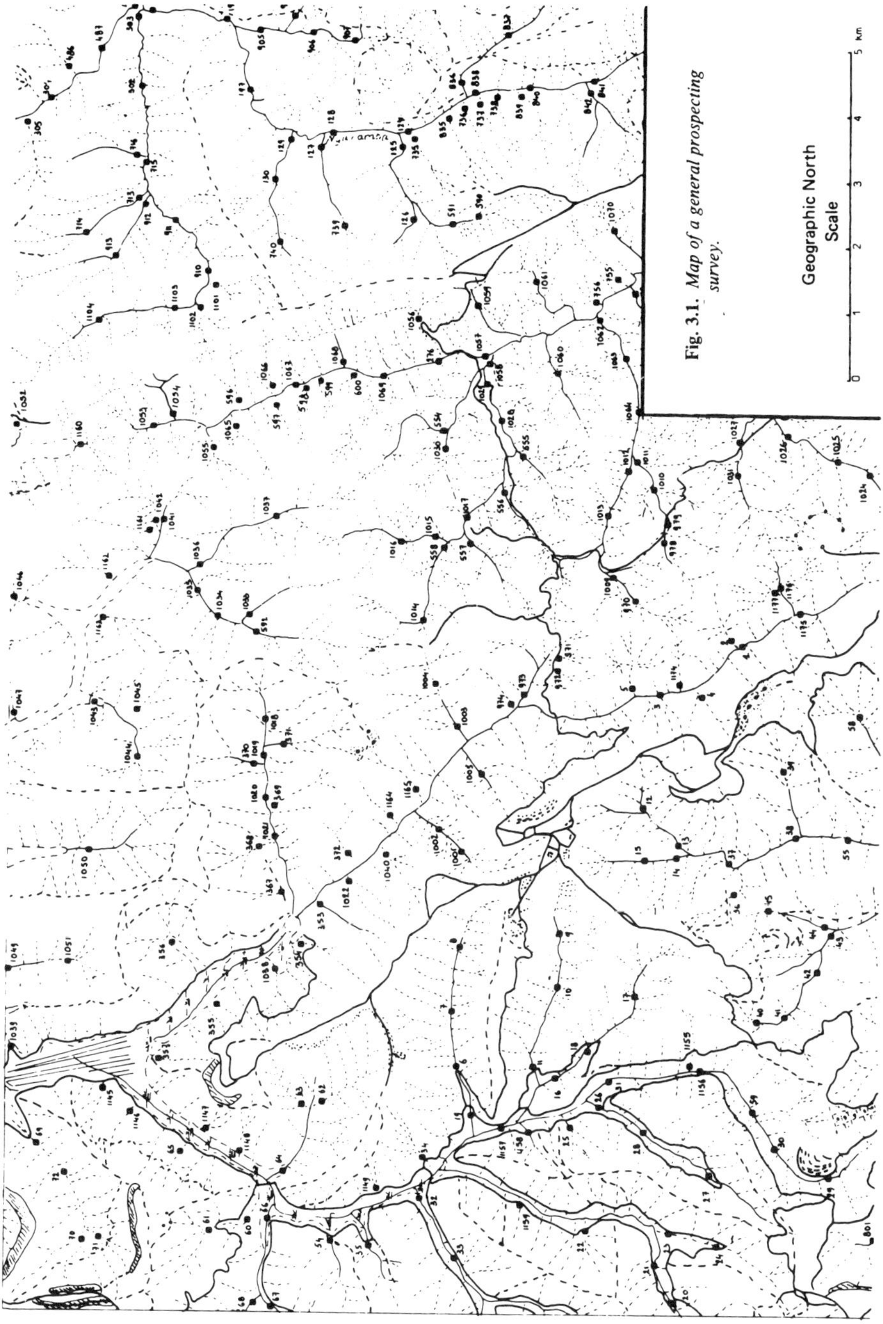

Fig. 3.1. *Map of a general prospecting survey.*

lateral), either separated or mixed according to the general survey guidelines.

1.2.4 SURVEY SCHEDULE

Set up a schedule for the traverses per group giving the number, distance and number of alluvial and geochemical samples to be taken.

1.2.5 DISTRIBUTION OF TRAVERSES PER GROUP

Each group leader makes a tracing of his traverse showing the positions of the alluvial and geochemical samples, and then reproduces it on to a 1:50,000 photogeological map. The geologist explains the traverse to the group leader and indicates on the map any specific points which require examination (contact, showing, hill for bauxite, etc., type of rock to be observed, etc.), and, if applicable, shows the samples and facies to be examined more closely.

1.2.6 CARRYING OUT THE TRAVERSES BY THE PROSPECTOR OR ASSISTANT GEOLOGIST

A good prospector can accomplish 8km (linear on map) and three or four alluvial samplings daily in an equatorial forest. The crew carries the camping equipment, prospecting instruments and rations. The group leader notes all his observations in the field log and on the map, ie position of samples, petrography, change of petrographic facies and tributaries. This map will be kept on completion of the field work and filed for subsequent use, perhaps to verify a disputed point in the drafting of a final survey report (Appendix 4, Chapter 1).

1.3 Active stream bed sampling technique

In temperate countries and in most tropical countries of uneven relief active stream bed gravel is readily accessible. Samples can also be taken from bank or terrace gravel beds.

1.3.1 SPACING OF SAMPLES

Spacing is usually 1km, sometimes 750 or 500m, apart determined before starting the survey in accordance with the geology of the region.

1.3.2 DENSITY PER SQUARE KILOMETRE

The density, one to five samples per km^2, is determined before the survey begins again according to the geology of the region.

1.3.3 SAMPLING TECHNIQUE

Each sampling procedure means that two samples are taken at 10 to 20m apart in optimal concentration sites for heavy minerals where the gravel is fairly thick and in as little mud as possible. The gravel is sampled at a depth of between 10 and 40cm, poured into a 5mm sieve placed above a pan, cleaned and screened until 5 litres of sand are obtained. The oversize is roughly classified by a few vertical shakes and dumped on the ground to check for the presence of large mineralized grains. The 5 litres of sand gathered in each pan are washed to obtain a concentrate containing not only heavy minerals but also a reasonable amount of average density minerals. The concentrates of both pans are combined in the same sample bag and labelled (eg by pan washing techniques and sample numbers).

1.3.4 ORGANIZATION

1) One prospector or group leader.
2) One or two samplers/washers.
3) Two reinforced edged shovels.
4) Two 5mm sieves (30cm diameter).
5) Two 8 litre stainless steel pans.
6) Metal bowls for waste.
7) Tubes and labels.

The yield varies according to the terrain. In France on average 12 samples are obtained in an 8h day, which includes washing but not drying of the concentrate. In tropical areas, considering the difficulties of forest coverage, this can be reduced to two to four samples for 8 to 10 linear km on the map or 1/2 samples per km^2, ie one alluvial sample for two geochemical samples in combination prospecting, which is the usual case.

1.4 Pit sampling technique

In certain peneplain regions in tropical areas sandy fills are extremely thick, to the point where the active stream beds remain sandy and poor in heavy minerals. Meanders are numerous and bedrock rarely visible. Prospecting via pits must be carried out in these areas, penetrating the ancient and often thick alluvial deposits resting on the bedrock. Pit prospecting is also especially recommended in searching for diamonds. To detect this mineral it is often necessary to wash 2 to 3m^3 of gravel at each sampling point.

2 General prospecting

The goal of general prospecting is to make a rapid check of mineralization showings found in a previous phase, in other words to make a rough determination of the extent and tenor. The procedures listed below are recommended out for this purpose:

1) Analysis of drainage pattern: length and breadth of waterways, flow regime, slope, discharge, alluviation.
2) Brief survey of secondary deposits: width of flats (flat spaces of varying size

along the waterways), thickness of overburden and gravel, nature of bedrock (substratum of secondary deposits, weathered to varying degrees), deposit dimensions, deposit type (beach, flat, terrace).
3) Topographical and geological observations.
4) Preliminary estimate of reserves in accordance with the average grades found (in gravel and excavated material).

Members of one prospecting team comprise:

1) 30 to 40 porters who perform unskilled labour in addition to porterage;
2) three to four group leaders;
3) one prospector.

The average estimated output is 2 to 3m^3 of gravel washed daily.

2.1 Location of sampling points

2.1.1 DETAILS OF SAMPLING POINTS

In the first instance sampling the surface sand cannot give any idea of river concentrations. Pits must be dug and this is he usual situation for main drainage courses **(Fig. 3.2).**

The ideal situation is where samples are taken at a convex bank and pits are drilled reaching the bedrock, if necessary, to scour the pot-holes (a pot-hole is a deep, steep-sided depression in the bedrock where heavy minerals accumulate as the

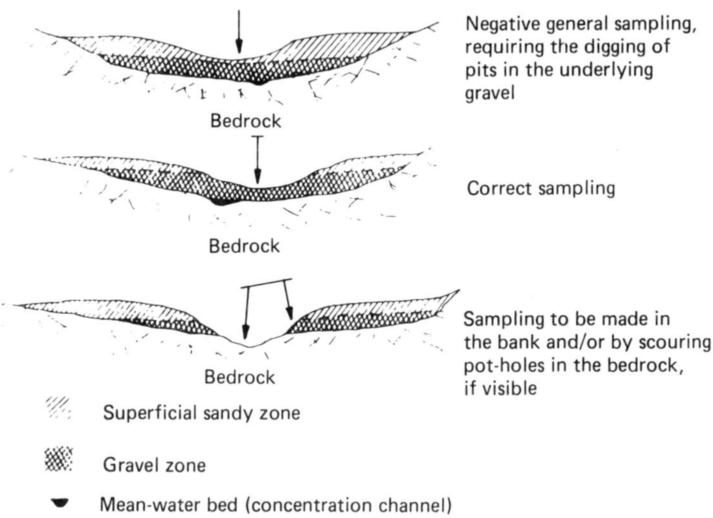

Various positions of the bed of a water course in relation to the different zones of an alluvial deposit

Fig. 3.2. *Position of sampling points.*

result of circular erosion). In this case a trench is often dug to the width of the bed and samples are taken at the point of contact with the bedrock.

The third procedure is when samples are taken from the banks, in the alluvium, fairly high above the level of the river. However, it is also important to take specific samples from any pot-holes in the bedrock which, despite reworkings, may indicate the qualitative content of the ancient alluvium. In addition, since this involves deep penetration of ancient alluvium, a sample taken from the concave bank in the steep side of these ancient alluvium is very important because it is representative of the ancient deposit **(Fig. 3.3)**.

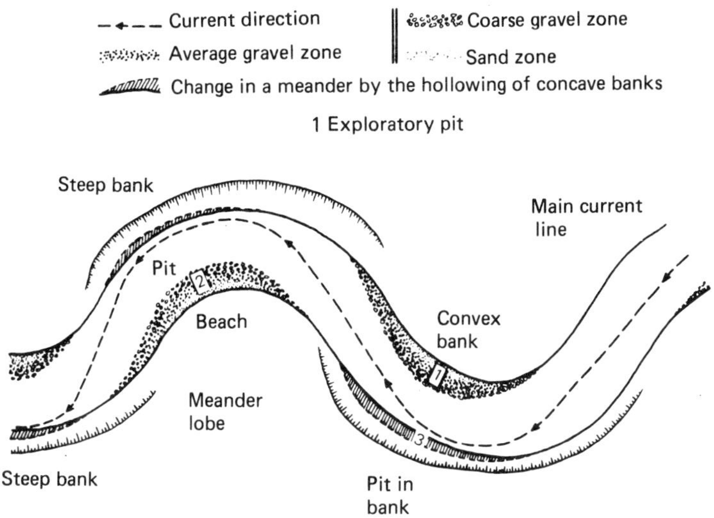

Fig. 3.3. *Lateral distribution of enriched zones in alluvial deposits (from M Derruau, 'Précis de Géomorphologie').*

In areas where there are rapids or falls it is useless to take samples from the pot-holes downstream because erosion is such that heavy minerals do not remain long and are quickly expelled. One sampling upstream of the rapids before reaching the eddies is often better. Depending on the type of bedrock or schistosity heavy minerals are retained to varying degrees, especially if the schistosity is inclined upstream and acts as a natural sluice.

When taking an alluvial sample on the edge of a trail do not forget to take samples at least 50m upstream from any bridge or raft, and never downstream, since the construction can alter the ratio of the minerals.

2.2 Pit sampling techniques

2.2.1 METHOD

Before deciding on the location for a pit, the presence of gravel at the point selected is generally confirmed by means of a drill rod which is embedded vertically into the ground and serves to guarantee the presence of a layer of gravel, its depth and its probable thickness. Prospecting begins by using isolated pits: one pit rectangular in section on each bank of the waterway under consideration, preferably in the convex part of two successive meanders. The major axis of the pit must be perpendicular to the river bed. If the river is no wider than a few metres and the gravel is sufficiently thick, it is acceptable to dig into the river bed itself (obviously in a dry part of the bed).

If required, a dam can be built upstream from the future pit and the course of the river diverted by a bypass of suitable length which can be used to feed the washing sluice. If sufficient time is available and if the alluvial deposits form flats, prospecting is continued using rows of pits, crossing each flat perpendicular to the valley axis. If a complete row proves to be non-productive, prospecting by isolated pits is resumed.

2.2.2 SPACING OF THE ROWS OF PITS

The space between each row should be about 400m, 200m for small waterways and 1000m if the mineralization is well distributed throughout the gravel. Along each row the pits should be spaced at 10 to 30m intervals. As as rule, one of the pits is placed near the mean-water bed.

2.2.3 SPACING OF ISOLATED PITS

Pits are placed approximately every 400 to 500m and this distance can be reduced by half for small waterways. It is generally decided according to the geology by the expedition leader.

2.2.4 SIZE OF PIT

A rectangular section pit **(Fig. 3.4)** should be dug, measuring 0.80 × 0.50m to a

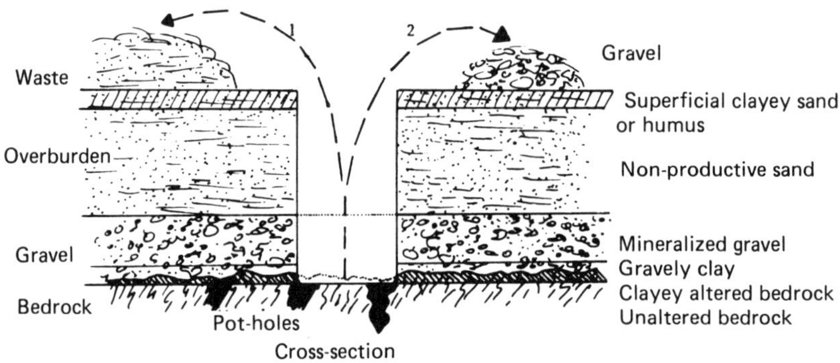

Fig. 3.4. *Cross-section of an exploratory pit.*

depth of about 0.60m; for widths between 0.60 and 3m in depth, cross-sections of 1.50 to 1.80m × 0.70 to 0.80m; for pits 3m or deeper, circular section 3.60 × 0.80m with one or two steps. If large boulders are present, the initial section (0.80 × 0.50m) must be increased to 1.20 × 1m. When prospecting for diamonds, which requires large volumes, the pit (or trench) cross-sections can, according to the thickness of gravel detected, reach 4 to 6m in length and 2 to 4m in width. Square section wells, with 5 to 6m sides, are frequently dug. The overburden is discharged to one side at a certain distance from the pit edge (danger of material falling back into the pit), with the gravel on the other side. The top 10 to 15cm of bedrock must also be collected and added to the gravel. The boulders must be washed in separate piles. If the gravel forms several superimposed layers, a drill rod should be used to make sure that the level assumed for the bedrock does not indicate 'false bedrock'. If the gravel is present in several superimposed layers alternating with clayey levels, it should be placed in separate piles since each layer may be mineralized differently or may be non-productive waste.

2.2.5 SUPERVISION OF DRILLING

A group leader supervises the digging of the pits and observes the safety regulations for the work force by ensuring that pit walls are vertical and that the cross-section of the pit from the surface to the bedrock is uniform. He must be able to take pit measurements correctly with a measuring tape (thickness of gravel and overburden)or failing this with previously cut rods. If necessary, he must also timber the walls if the soil does not hold (Chapter 7).

2.2.6 DISPOSAL OF CUTTINGS

If the pit is more than 2m deep, one worker at the surface removes the cuttings in a bucket attached to the end of a rope. The bucket is filled by the digger who works at the bottom of the pit.

2.2.7 PIT DRAINAGE USING BUCKETS

The water must be dumped on the naturally sloping side of the land to ensure that it does not reenter the pit by seepage. If water seepages are significant, dewatering can be accomplished by means of a pump with a capacity suitable to the inflow of water (Chapter 7).

2.2.8 SAFETY INSTRUCTIONS FOR SINKING A PIT IN A NON-CAVE FORMATION

For pits more than 2m deep the pit sinkers must work in groups of two. In the case of an accident the worker remaining at the surface can seek help. They must wear hard hats and must be attached to a rope to ascend and descend. The rope, which must be properly fastened to the belt of the shoveller who descends, is held by his assistant who winds it around a post firmly embedded in the ground. The assistant shoveller stands back from the edge and allows the rope to slide down as requested by the shoveller. To prevent soil from being dragged down into the pit where the rope slides a piece of wood is wedged into a channel across the corner of the pit.

2.2.8.1 Safety landings

Safety landings **(Fig. 3.5)** must be provided for pits more than 5m deep. The first

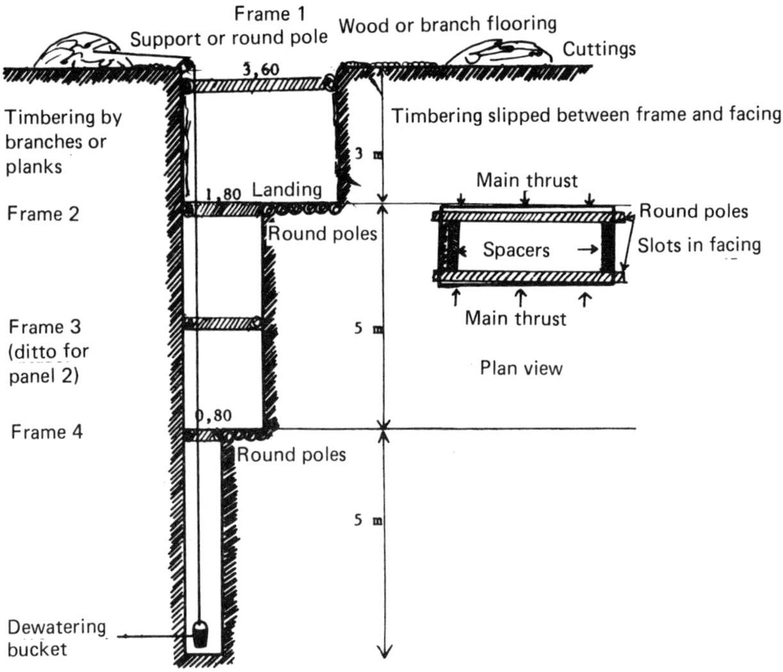

Fig. 3.5. *Section of an exploratory pit with timbering.*

landing must be located 3m from the surface, and then at 5m intervals. Protective flooring must cover at least one-third of the pit section. The flooring should consist of three round poles (minimum diameter 8cm) placed horizontally and embedded in the pit walls. The poles can be covered with bamboo or palm leaf stems and a pad of leaves or grass. Note that if the ground is good and does not require timbering, it is preferable to dig circular pits, since the soil holds better.

2.2.8.2 Flexible or rigid ladders

For pits more than 10m deep the worker can use a flexible ladder or a fixed metal ladder to climb out of or into the pit. The ladder must be fixed by its hooks to a round pole placed above the pit. The ends of the pole are placed in previously cut channels.

2.2.8.3 Summary of safety measures

1) Slots in the walls	0 to 2m
2) Hard hats mandatory	from 2m
3) Safety ropes	from 3m
4) Safety landings	from 5m
5) Flexible or metal ladders	below a depth of 10m

As a rule, these measures can only be observed in pits with very sound walls (eluvial or dry deposits in the Sahel zone). If the soil strength is poor, it is necessary to timber the pit walls or to sink the pit using a 'Banka' drill.

2.2.8.4 Equipment tests

Knots must be tied regularly in the extraction rope to prevent the rope slipping through the hands of the puller. The ropes must be in good condition and, if possible, of rot-proof nylon. The ropes should be gathered together at the end of the shift to be dried. Damaged ladder rungs must be replaced immediately. Buckets used to extract the water, and whose lugs and handles are weak, must be strengthened by running the rope under the bottom. Any object or tool lowered into a pit must be attached in duplicate to a rope or lowered in a bucket if its size permits. The surface periphery of the pit must be cleared for at least 1.50m and fine waste gravel spread around the opening to prevent sliding. No tool or other unattached object must be allowed to lie loose near the edge of the pit. Fires must not be lit near a pit (carbon dioxide is heavier than air and is liable to descend into the pit and asphyxiate the workers). It is wise to provide a ventilation system in certain cases, and to check that the air is breathable at the bottom of the pit, before descending in the morning (lower a candle or throw in a firebrand).

2.2.8.5 Pit protection in inhabited areas

An unfilled pit must be surrounded by a fence (four posts solidly embedded in the ground 1.50m apart to form a square and connected together by barbed wire at a height of 1m above the ground; if wire is not available, wooden poles can be nailed to the posts to form a barrier). The authorities of the neighbouring villages must be told of the risks to inhabitants and a 'DANGER' sign must be put up near the pit. If the drilling site is large, it must be fenced off completely.

2.2.8.6 Filling in the pits

In inhabited areas when pits are not longer being worked they must be filled in. In no circumstances should extracted gravel be used for this purpose. The top part of the pit should be widened into a funnel shape and the cuttings obtained used to fill in the lower part of the pit.

2.3 Drill hole sampling techniques

2.3.1 LIGHT-WEIGHT 'BANKA' DRILL

Since the data generated by pits is usually complete, drill holes should only be used when pits cannot be excavated. This applies to flats, which continue into the active stream bed of the river (drills mounted on pontoons), and flats where the overburden consists of moving sand. The drilling equipment generally used is the 'Banka' 4 or 6 inch drill. The very light 4 inch drill is used for strategic prospecting, whereas the 6 inch drill is better for deposit assessment, since larger samples can be obtained **(Fig. 3.6)**. The best operating conditions are obtained when the deposit is quite extensive and the rolled gravel with small uncemented elements contains very few boulders, and mineralization is fairly uniform. The bedrock must not be too hard.

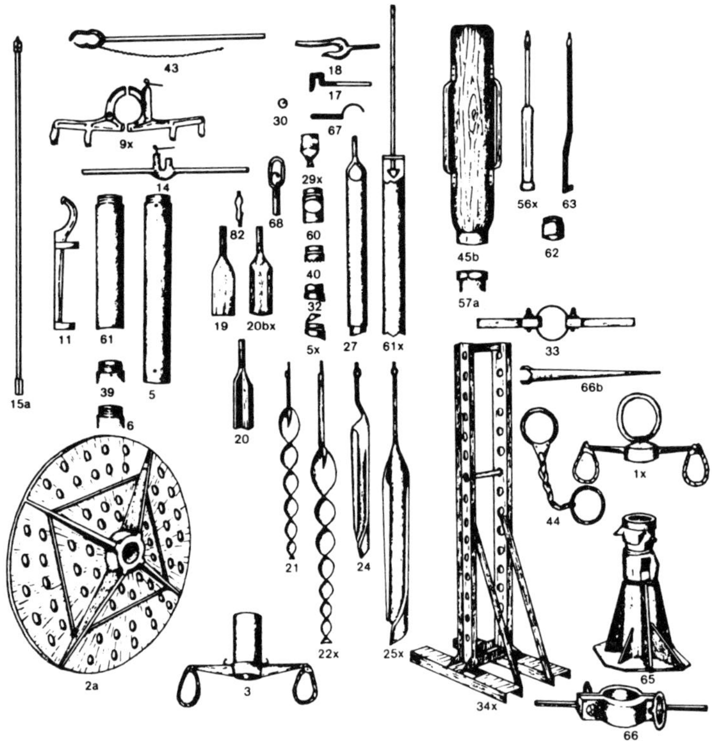

Tool bit			Tool bit		
4"	6"		4"	6"	
1	1a	Pipe extractor cap	32	32a	Shoe for sand pump with ball valve
2a	2a	Sectioned platform	33	33a	Support tongs
3	3a	Platform coupling	34	34	Derrick
5d	5g	Cover pipe	39	39a	Toothed shoe for elongation pipe
6	6a	Short shoe for elongation pipe	40	40a	Toothed shoe for sand pump
6f	6g	Long shoe for pipes	43	43a	Chain tongs for pipes
9	10	Pipe clamp	44	44a	Chain link to remove pipes
11	12	Pipe dog	45b	45b	Wood driver
14	14	Rod clamp	54	54a	Flat disc shoe for sand pump
15a	15a	Drill rod	56	56	Slide
17	17	Lifting wrench	57a	57a	Bumper to tamp pipes
18	18	Catch wrench (fork)	60	60a	Bit shoe for sand pump
19	19a	Flat bit	61	61a	Piston type bailing pump
20	20a	Star bit	62	62a	Stone catcher sleeve
20b	20c	Ear bit	63	63a	Hook
21	22	Small auger	65	65a	Hollow jack
22	23	Large auger	66	66a	Pipe clamp
24	25	Small gouge	66b	66b	Wrench for clamp knobs
25	26	Large gouge	67	67a	Wrench for pump shoes
27	27a	Sand pump	68	68	Swivel stem
29	29a	Auger shoe	82	82	Bit coupling
30	30a	Ball valve			
– 6 m long collapsible steel derrick					
– Idler pulley					

Fig. 3.6. *Parts of the 'Banka' 4 and 6 inch drills.*

2.3.1.1 Personnel

Five to nine men are required plus a shift foreman, depending on whether drilling takes place in the overburden or in the gravel. The shift foreman fills in the drilling log and measures the depths reached, either using rods or pipes. He must ensure that the measurement of the drill string does not exceed the length of the casing tube, to avoid jamming the recovery tools under the shoe of the first drill pipe and excavating a wider hole below the casing.

These operations, albeit simple, nevertheless require the use of a trustworthy trained operator.

2.3.1.2 Average daily penetration levels (per 8h shift)

1) In sandy overburden: 30 to 40m.
2) In clayey overburden: 10 to 30m.
3) In gravel: 1 to 15m depending on the need for a drill bit, whether the gravel is sandy, loose or clayey.

2.3.1.3 Small auxiliary equipment and sample bags

A large number of plastic bags (30 × 40cm) are used to recover the samples. Once the water-table has been crossed, if the gravel is loose, material is frequently raised through the pipe and a large number of sample bags is necessary. The average number of sample bags used by a 'Banka' 4 inch drill is about five bags per metre drilled. Hence resources must be put aside for sample bags as well as for small auxiliary equipment.

2.3.1.4 Advantages

Ruggedness: the drill is built of steel and will withstand great impact. The drill, which can be dismantled into parts weighing less than 30kg, allows installation in places that are relatively inaccessible to other types of drill. The 'Banka' drill is simple to handle and quickly mastered by the unskilled workforce. It allows drilling below the water-table, to more than 40m, in sand/gravel formations of highly variable grain size.

2.3.1.5 Drawbacks

The weight and the number of drill parts mean that the complete system must be transported by truck if the drilling schedule is to be maintained.

Errors can be introduced when taking measurements: simple operation but nevertheless demands constant attention throughout the drilling operation, which must be conducted by a literate conscientious shift foreman.

Pipe driving and lowering operations must be synchronized, pipe penetration always preceding that of the drill string bearing the recovery tools, particularly to prevent the 'pump' from continuing to extract gravel when the casing has not reached the hole bottom. Some of the extracted gravel could in this case be taken from outside the theoretical cylinder formed by this casing.

A large number of minor incidents can hamper penetration: jamming of tools or detachment in the casing, breakage of equipment which must be rewelded, hence the need to have an electric welding set permanently on the drilling site. A large strong workforce must be available for drilling, and additional labour is needed to treat the samples on the spot. Despite the advantages of the 'Banka' drill, its main drawback is its slowness.

2.3.2 MECHANIZED DRILLS

For greater efficiency mechanized drills such as the Meca-Banka, the drilling tractor or the Benoto drill are used. In assessing deposits the latter two offer the advantage of making large diameter holes (625mm for the well tractor), allowing the retrieval of large gravel samples which are required when mineralization is not uniform or if it is in the form of nuggets.

2.4 Gravel sampling and washing

2.4.1 GRAVEL SAMPLING

Several processes are available to sample gravel from one pit:

1) Fill three pans with gravel from the gravel extracted from the pit, after mixing it well on the surface. This constitutes a standard average sample.
2) Fill two to four pans with gravel from the entire thickness of the gravel in place, in the pit, using the vertical channel cut out of two opposite facings or four facings of the pit, making a total of two to four channels.

2.4.2 WASHING AND CONCENTRATION

Gravel samples taken in the active stream bed or from a pit are washed in a buddle or pan, using a hand jig or sluice.

2.4.2.1 Buddle or pan washing

The samples are washed where they are taken wherever possible, or at a washing station set up at the centre of the prospected area, on the bank of a readily accessible water course or failing this in drums or in artificial basins in an arid zone. The gravel is cleaned of mud on a 5mm mesh screen, placed above the water or a bowl. This operation must be performed very carefully. The oversize is rapidly examined; in the case of visible mineralization the samples are stored. Screenings under 5mm are cleansed in a pan or bowl and the mud is carefully poured off.

Heavy minerals are then separated by means of the buddle or pan. The conical buddle, generally metallic, requires a fairly large water surface for correct handling. The instrument, which floats on the water, is subject to tilting motion, so that the materials rotate, are separated by the centrifugal force and rise to the top at a point opposite the operator. If the washer works in the streams, the light elements are removed in the direction of flow. Products which sometimes tend to collect at the centre of the buddle can be removed by slight horizontal shakes, which a highly experienced panner can manage while 'rotating' the buddle very gently to spread the minerals.

The pan, in the form of a truncated cone, does not rest on the water. It is held with both hands, with the operators elbows pressing against his legs just above the knees. The operation includes cycles of three successive movements, repeated until the light minerals are eliminated:

1) horizontal shaking, identical to that of screening, with the pan moderately filled with water and kept horizontal;

2) tilting of the pan towards the side opposite the operator and spillage of the water without losing any sand;
3) while maintaining this tilted position, immersion of the pan in the water, stream or bowl to add water and remove it immediately, so that the upper layer of sand is reduced (movement to be repeated a maximum of three times);
4) resumption of the horizontal position and shaking, etc. and the entire procedure is repeated.

Pan washing offers certain advantages over the buddle:

1) low water requirements (a bowl can suffice);
2) better recovery;
3) possibility of recycling waste when the pan is used above a recovery tank.

The BRGM has developed a light, compact manual prospecting jig which can concentrate rapidly heavy minerals larger than 0.5mm. This tool can replace the buddle or pan or be used together with them for particle sizes over 1mm. If large quantities of gravel have to be washed, to estimate the grade for example, and the substance prospected (especially gold) is in nugget form, it is necessary to use a sluice **(Fig. 3.7)**.

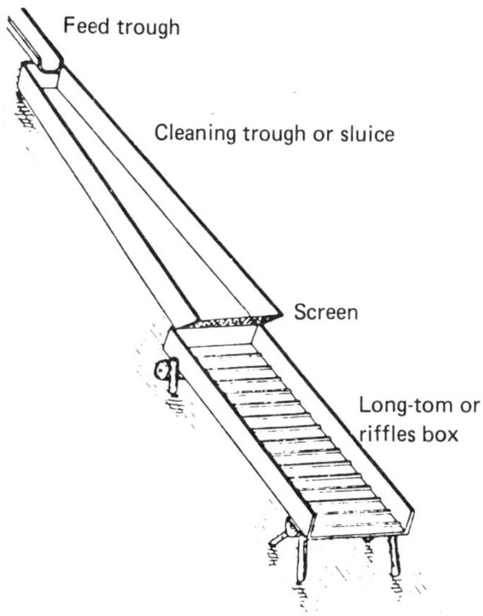

Characteristics

1 Cleaning trough:
Length: 3.5 to 4 m
Width of lower part: 0.70 to 0.80 m
Width of upper part: 0.40 to 0.50 m
Height: 0.20 to 0.25 m
Slope: 8 to 10%

2 Screen:
Inclination to floor: 45°
Diameter of opening 1 to 1.5 cm

3 Long-tom or riffles box:
Length: 2.5 to 3 m
Width: 0.90 m
Height: 0.20 to 0.25 m
Slope: 6 to 8%

Fig. 3.7. *Diagram of a long-tom short sluice.*

2.5 Analysis of heavy minerals in a concentrate

In general prospecting the term concentrate is applied to the 'black minerals' left in the buddle after the removal of minerals forming the white border. However, in the very rare case of diamond prospecting by buddle, and in any case, it is always advisable to examine the buddle contents carefully on a screen before reaching the black minerals. Do not trust the old gold prospectors who tend to push the concentrates beyond the black border, only keeping the grains of gold dust and the 'colour'. After concentration the edge of the concentrate and the buddle bottoms are examined for an initial rapid determination before drying and packing **(Table 3.1)**.

To describe the buddle washing result, the following terms are recommended for gold:

1) colour: gold particles so fine that they are only recognizable under the binocular microscope, or by their overall colour, if large amounts are present;
2) points: slightly larger particles, yet not large enough to be identifiable under a magnifying glass;
3) small grains: gold particles with an apparent surface but with a thickness not identifiable under a magnifying glass; they can be counted under the binocular microscope;
4) large grains: gold particles with three clear dimensions; these are called 'small' if their dimensions do not exceed 1mm, and 'large' if their dimensions lie between 1 and 2mm;
5) nuggets: fragments larger than 2mm in size.

Note that gold is highly mobile under certain conditions. It is liable to travel in solution quite far from its original deposit and to be fixed in the presence of humic acids, implying that nuggets may be present several kilometres downstream from the source and disappear as the latter is approached.

2.5.1 COLOUR/CONTENT EQUIVALENCE

For gold and cinnabar, the BRGM uses an empirical method to quantify traces based on the measurement of the average diameter of the grains or 'colours'. Each particle is pushed by a needle on a scale graduated in half-millimetres engraved on the bottom of the observation dish. The average diameter is estimated in tenths of a millimetre and the diameters of the different grains are added together. Correspondence is approximately the following for grains not exceeding 5/10mm and giving an average diameter of 3/10mm:

1) gold: 500 tenths of a millimetre equivalent to 0.1g;
2) cinnabar: 900 tenths of a millimetre equivalent to 0.1g.

The grains or small nuggets exceeding half a millimetre are hypothetically broken down into grains measuring 0.2 or 0.3mm, or, if numerous, they are weighed directly. Amalgamation is sometimes necessary for gold separation and weighing. The method of evaluation in tenths of a millimetre provides an order of magnitude of the content per cubic metre of alluvium. For example, if 10 litres of washed sands give 55/10mm of gold, the gold content recoverable by buddling is estimated to be $1 g/m^3$.

Scale of hardness (or MOHS scale).

The mineral represents one degree above the one preceeding it:

1 talc	3 calcite	5 apatite	7 quartz	9 corundum
2 gypsum	4 fluorite	6 feldspar	8 topaz	10 diamond

These figures are valid for unaltered minerals

Table. 3.1. *Commonest alluvial minerals. Panning classification.*

Mineral	Characteristic colours	Form	Observations
1. *Minerals floating on surface, sometimes forming film*			
Graphite	Grey, shiny	Flakes	remain on surface
Micas	White, yellow, black, brilliant	Flakes	
Molybdenite	Light grey, brilliant	Flakes	particularly after crushing
Specularite	Brilliant, black	Fine flakes	
Sulphides (pyrite, chalcopyrite, pyrrhotite. etc.)	Grey, yellow, metallic	Dust	
Gold	Metallic, golden yellow	Fine flakes	
2. *Minerals forming white ring*			
Quartz	White, grey mat or shiny	Grains, often fragments with crystalline faces	Largest are included in next category
Feldspar	White, pale yellow, pink	Grains, often cleavage fragments	
Calcite	White, pale yellow grey	Grains, often cleavage fragments	
Beryl	White, pale yellow, emerald green	Grains, fragments of hexagonal prisms	
3. *Outside part of black tongue (black sand)*			
Fluorite	Violet, green	Grains	Largest identical with next category
Andalusite	White, greyish	Fine elongated prisms	
Apatite	Grey, light pinky-brown	Fragments of hexagonal prisms	
Tourmaline	Black, dark green	Fragments of prisms with triangular section straited longitudinally	
Amphiboles	Dirty green, black green	Grains, cleavage 124°	
Pyroxenes	Dirty green, black green	Grains, cleavagee 90° approx	
Pisolites	Rust brown	Grains, more or less spherical	
4. *Inside part of black tongue (black sand)*			
Ceylonite	Green	Grains, of ten octahedron	Maybe attracted to magnet Particular refraction in water
Topaz	Colourless, light green	Elongated, striated prisms	
Kyanite	Pearly white, shimmering blue	Fragments, flattened prisms	
Corundum	Light yellow, light grey	Grains	
Sphene	Pinky grey	Grains	
Epidote	Yellowy green	Grains, fragments of prisms	
Staurolite	Dirty brown	Fragments, orthorhombic prisms	
Olivine	olive green	Grains, rounded	
Garnets	White, mauve, red-brown	Grains, spherical with many faces	
Specularite	brilliant black	Grains, flake shaped	
Ilmenite	Purpley-black	Grains, flake shaped	
Magnetite	Brownish black	Grains, octahedrons	
Diamond	White, grey	Octahedrons, sometimes with curved edges, cubes	
Zircon	Grey, colourless, yellow, red	Grains, tetragonal prisms	
Rutile	Metallic grey brown	Fragments, prisms with cleavages	
Scheelite	Yellowish grey	Grains	
Monazite	Dull orange brown	Fragments, fine prisms	
5. *Bottom of pan*			
Wolfram	Black	Grains, fragments, prisms	Constant intensity of colour and shine when lit from any angle*
Chromite	Black	Grains, octahedrons	
Cassiterite	Yellow, dark brown, black	Grains, sometimes cleavages	
Columbo-Tantalite	Dull, dark brown	Grains, sometimes cleavages	
Thorianite	Black	Grains, cubes	
Native copper	Pink-red with metallic shine	Grains, specks, flakes	
Gold	Gold-yellow with metallic shine	Grains, spots	
Platinum	Grey with metallic shine	Grains, spots	

* Detail allowing native gold to be distinguished from some metallic sulphides, particularly of chalcopyrite, the grains of which are not shiny from some directions.

It is useful to measure the particle size distribution of the buddle bottoms using a particle size pen such as the one developed by the BRGM.

2.6 *Sample labelling and packing*

2.6.1 SAMPLE DRYING

During sampling and after pan washing and concentration the concentrate is temporarily stored in a tight plastic bag, carefully sealed and labelled, after having rinsed the buddle thoroughly with water to make sure that none of the recovered material is lost. The water in the plastic bag is then drained, with due care to avoid losing any concentrate.

This concentrate must be dried:

1) The same evening at the bivouac, or on return to the base camp at the end of the itinerary. In this case drying is carried out in a buddle over a wood fire (or on a gas heater over a gentle flame) to avoid altering the concentrates.
   Heating must not be too intense to avoid eliminating certain minerals by decrepitation or sublimation for example, or by the addition to the concentrate of metal splinters dislodged from the pan. To achieve smooth heating two pans are superimposed, separated by a few pebbles. The buddles must also be cleaned thoroughly after each drying procedure, and scoured by brush. It is essential to supervise the drying team to make sure that persons who are accustomed to grinding and screening geochemical concentrates do not afford the same treatment to alluvial concentrates.
2) If the concentrate can be conveyed every evening or every week to the laboratory, it can be dried in the oven, weighed and prepared directly for analysis.

2.6.2 PACKING THE CONCENTRATE

The dry concentrate is divided into two equal fractions, which are placed in two plastic tubes bearing the same number and the letters A and B. The first tube is sent to the laboratory (with the understanding that it represents only half of the total weight sampled) and the second is stored for use in case of discrepancies or if the original is lost. If the concentrate contains nuggets, they are weighed, packed separately and their weight entered on the label of concentrate A and B.

In this specific case (original lost) the remaining concentrate is again divided into two parts, so that only a quarter is sent to the laboratory and one quarter remains in storage.

2.6.3 SERIAL NUMBERING

One and only one number, assigned at the time when the samplings are collected, is used during sample analysis, treatment and data storage. A prenumbered field log with detachable sheets helps to reduce duplicate numbering errors to the minimum. This log can also be used to record field observations made at the time of sampling, in coded or some other form. In all cases of strategic prospecting and even for tactical prospecting serial numbering of the following type is employed:

Example: GU/19 0725
GU = Strategic prospecting in Guyana
19 = Zone 19 (delimited at the start of the prospecting programme)
0725 = Four-digit serial number

This numbering system offers the advantage of avoiding double or triple numbering, a frequent source of error. It is easy to use and can be applied at any stage of the prospecting operation. However, it is not intended to localize the sample, this information being provided by the coordinates, which must automatically accompany the sample identification in this system.

Identification is limited to four digits to avoid encumbering subsequent treatment by the handling of excessively bulky files. This necessity to have less than 10,000 samplings per zone can be circumvented easily by replacing the first two digits by a letter during treatment (11,000 = A;000, 12,000 = B;000). This system can also be employed to number tactical samplings. During a given prospecting programme involving different groups the numbering system using initials is discarded but each group is given a set of numbers (Group 1: 0 to 0999, Group 2: 1000 to 1999, etc.). From a practical standpoint the sampling group leader at the BRGM leaves every morning with his prenumbered pad and a series of prenumbered labels which are attached to the sampling (yellow plastic labels with the number written in indelible ink or ballpoint). Entries made in advance at the camp are normally more legible than entries made in the field.

2.6.4 SHIPMENT

Packing the tubes for shipment, often for a road journey followed by air freight, must be carried out with extreme care. A reliable solution, one that appears perfectly satisfactory, involves the use of cardboard sheets, perforated to match the diameter of the tubes used. Each sheet is planned for a fixed number of tubes: 80 large tubes or 120 small tubes. The safest solution is to have special boxes made, in which an exact number of sheets can be stacked.

Spaces that are left empty are liable to cause the samples immediately above or below to be lost. For greater safety insert empty tubes into these spaces. If metal chests are used, not corresponding exactly in size to the cardboard sheets, make sure all the samples are securely wedged with paper.

The numbers of the survey and of the samples are entered on each sheet. Example: GU 19-0001 to 0080; the box contains GU 19-0001 to 0400. Place a dispatch note in the box and a request for analysis, which must also be sent separately by mail. The BRGM uses another tried and tested system, which consists of packing the tubes in groups of 12 in a flat plastic pouch, laying 10 pouches flat in a large plastic bag and the bag sealed by adhesive tape or clips, identified by numbers on the tube labels, plus a sheet in each pouch of 12 samples and a second summary in the 10-pouch pack. A summary of the contents is added to the chest and a duplicate of the dispatch note sent to the laboratory by mail. This method is the safest for samples transported by truck over poor terrain or long distances.

3 Systematic prospecting and evaluation of alluvial deposits in streams and rivers

Only regions where general prospecting has identified mean grades above the workability limit are worth investigating by systematic prospecting.

3.1 Method of working

The sector to be analysed is a restricted area and must be examined in detail with a much finer grid than in the previous phase. Since questions of group mobility are no longer relevant at this stage of exploration, it is necessary to establish fixed camps not far from prospecting sites (no more than a few kilometres away, possibly with the use of a car track for personnel transport).

To simplify the calculation of reserves the survey region is covered by a prospecting network consisting of a succession of rectilinear units similar to the pattern employed in tactical geochemical prospecting **(Fig. 3.8)**. A baseline is created first, which then serves as a prospecting trail. It follows the general axis of the valley, remaining parallel to the main trends, and is finalized by a precise topographical survey. From this baseline prospecting (or pit) rows are drawn perpendicular to each side. The starting point of each row is marked by a stake taken as the origin for measuring distances on that row. These rows are parallel.

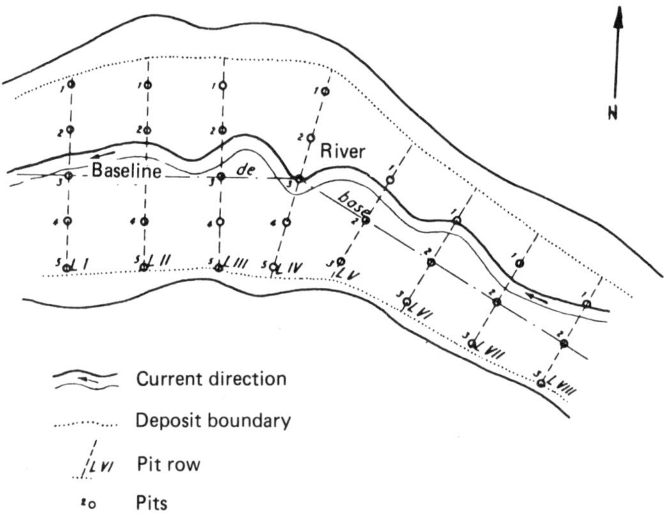

Fig. 3.8. *Plotting of pit rows (from R Bonte 'L'etude des gîtes secondaires métaux et pierres précieuses. Prospection et exploitation des alluvions').*

It may happen that the width of the flats is such as to make it impossible to determine the axis of the flat and the orientation of the banks at first glance, especially in a forested area. In this case it is first necessary to clear the area at the foot of the banks and survey the boundaries of the flat, then to plot the theoretical axis and place the row on the map and then on the field. The rows must stretch from one edge of the flat to the other. They are designated by a letter of the alphabet or by figures preceded by the letter R (rows). The pits (or drill holes) are placed at equal distances along each row. They are numbered in Arabic numerals, from left to right up the valley. The spacing of the rows and of the pits on each row is specified by the geologist. Depending on the size of the flats, and the uniformity of the deposits and mineralization, the following spacing can be selected:

1) between pit rows: 400, 200, 100, 50 or 25m;
2) between pits: 20, 10 or 5m.

The distance of 400m between the rows of pits must be maintained in non-productive zones and is reduced to 200m as soon as grades equal to or greater than the minimum grade are encountered.

As a rule, the density of the pits per unit area varies according to the deposit conditions of the substance, and the accuracy desired. Consequently, the more irregular the mineralization, the greater the density. 'Banka' hand drills could be used for swampy flats.

3.1.1 SAMPLING AND VOLUME OF SAMPLES

Sampling is achieved by making vertical cuts through the entire gravel thickness, two parallel facings or on the four facings of the pit or by taking an average sample from all the gravel extracted from the pit after mixing. Pebbles larger than a fist (boulders) are set aside as the gravel is extracted. Their volume must be measured and compared with that of the boulder-free gravel. This calculation establishes the percentage of boulders which will subsequently be taken into consideration in calculating the ore reserves visually. The operation is performed as follows: all the boulders contained in a given pile of gravel are separated and a cake is made with the rest of the gravel. Two rods are cut of the same length as the diameter of the cake and placed in a cross shape next to the gravel. A circle is then drawn on the ground, with the two crossed rods for perpendicular diameters. The quadrants of the circle are then filled with boulders so that the height is uniform throughout **(Fig. 3.9)**.

If H1 is the height of the gravel cake, H2 that of the pile of boulders and these boulders only fill a fraction f of the circle, then the proportion of boulders in the gravel is:

$$p\ \% = 100 \times \left(\frac{H_2 \times F}{H_1 + H_2 \times F} \right)$$

Example: H1 = 60 cm
H2 = 20 cm
f = 1.5/4, or 0.375 (one and a half quadrants)

$$p\ \% = 100 \frac{20 \times 0.375}{60 + (20 \times 0.375)} = \frac{750}{60 + 7.50} = \text{about } 11\%$$

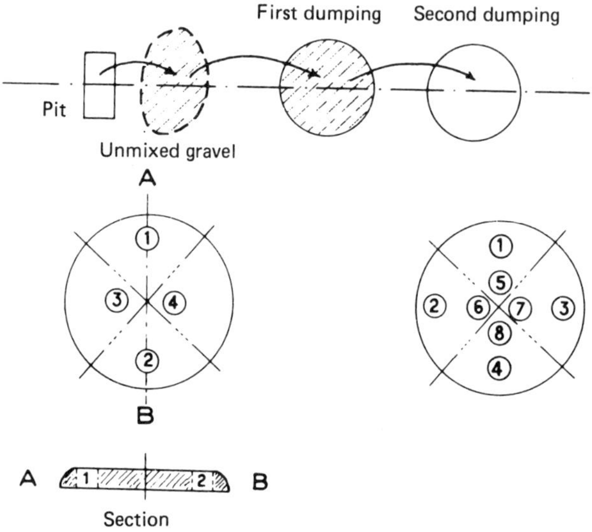

Fig. 3.9. *Taking samples from extracted and mixed gravel.*

The gravel is mixed by shovel, spread in uniform piles, and then dumped completely twice in succession. This is carried out by two labourers working simultaneously. The second dumping can only begin after the first is completed, and the first not before the pit is fully excavated. During the mixing make sure that the gravel does not contain any branches or dead leaves, or as little as possible. The piles must be perfectly flat and must not be more than 50 to 60cm high. They must not be packed either on the upper surface or on the sides. The sample is taken after this double dumping. The pans are taken as follows:

1) four for thicknesses less than 2m;
2) eight for thicknesses greater than 2m.

3.2 Calculation of content

3.2.1 ESTIMATE OF CONTENT IN GRAVEL (OR IN PLACE)

$$t = \frac{p \times N}{n} \quad \text{or} \quad t = \frac{p \times V \times K}{v}$$

where:

t is the content in g (or carats) per m^3;
p is the weight in g (or carats) collected in the washed sample;
n is the number of pans washed;
N is the number of pans per m^3 (150 8-litre pans for the increase in volume);
v is the quantity washed (in litres);
V is the reference volume, 1000 litres (1 m^3);
K is the expansion factor or number of litres of expanded gravel corresponding to 1 litre of gravel in place.

3.2.2 DETERMINATION OF THE VOLUME OF GRAVEL IN A PIT

To determine the volume of gravel in a pit, if the gravel is fairly fine and the pit uniform, it is enough to measure the areas of the well at the top and bottom of the gravel. The volume formed by the truncated pyramid is then calculated:

$$V = \frac{x}{3}(Ss + Si + \sqrt{Ss \times Si})$$

where:

V is the volume of gravel;
x is the gravel thickness;
Ss is the area of pit at the top of the gravel;
Si is the area of pit at the bottom of the gravel.

If the pit walls are not uniform, especially due to the presence of large boulders, the volume V must sometimes be evaluated more accurately by a breakdown into elementary volumes.

3.2.3 ESTIMATE OF CONTENT PER SQUARE METRE OF GRAVEL

$$tc = t \times g$$

where:

tc is the content in g (or carats) per m^2 of gravel;
t is the content in g (or carats) per m^3;
g is the gravel thickness.
1 carat = 0.2g.

3.2.4 ESTIMATE OF CONTENT PER CUBIC METRE EXCAVATED

$$te = \frac{t \times g}{g + S} = \frac{t \times g}{X} \quad \text{or} \quad \frac{tc}{g + S} = \frac{tc}{X}$$

where:

te is the content in excavated material;
t is the content in gravel;
tc is the content per m^2
g is the gravel thickness;
s is the thickness of non-productive material;
X is the total thickness of the alluvium.

In the presence of nuggets or grains of abnormal size these elements must be set aside and weighed separately. The contents can then be estimated either by taking account of their weight or by ignoring them. It is important to proceed always in the same manner, either one way or the other, and the observations column must mention the tables identified, with the weight of the exceptional elements weighed separately.

The data must be complete. All observation made in the field must be entered in the prospecting log **(Table 3.2)**. The figures and all the data in this log will subsequently have to cube the non-productive material and gravel in a deposit, to determine the content per m^3 of gravel, per m^2 and m^3 of excavated material, as well as the quantity of ore contained. Moreover, all the useful information must appear on the corresponding prospecting maps: names of rivers, traverses and roads surveyed, plot of numbered rows, location of works with the corresponding number, thickness of non-productive materials, gravel thickness, mean contents, etc.

As a rule, a scale of 1:1000 should be adopted for each river, 1:5000 for the water course and 1:10,000 for the entire area. Naturally, the maps must be oriented to magnetic North, identified by the date of their preparation and geographic situation, and have a complete key (arrow pointing to magnetic North, scale in figures, name of sector investigated, prospector's name, signature, etc.). In some cases (eg in diamond prospecting) it is wise to have accurate maps (to scale 1:10,000 or 1:20,000), reconstructing the entire hydrographic history of the sector, after a survey of terraces and old beds, or specifying the location and composition of minerals accompanying diamond, characteristic for a given deposit, etc.

3.3 *Assessment of alluvial deposits*

This is done to estimate the actual reserves contained in the deposits. It precedes and, in effect, constitutes the first phase of mining. Calculations are generally made at the end of prospecting but the prospector, according to his prospecting maps and index cards, must himself be capable of estimating the approximate total value of the deposit that he has prospected. The following measurements can be calculated:

1) length of the deposit;
2) mean breadth of the deposit (the arithmetic mean of the different breadths of the deposit considered);
3) mean thickness of gravel (also the mean of gravel heights in the prospecting pits);
4) mean thickness of non-productive material (the mean of the heights of the non-productive material in the prospecting pits).

This gives:

Volume of gravel = Sg and Volume of non-productive material Ss

where:

S is the total area of the deposit;
g is the mean gravel thickness;
s is the mean thickness of non-productive material.

Hence the mean contents in the gravel for the overall deposit are:

$$\text{Tm/gr} = \frac{S_1g_1t_1 + S_2g_2t_2 + \text{etc.}}{S_1g_1 + S_2g_2 + \text{etc.}}$$

The mean contents in the excavated material are:

$$Tm/e = \frac{S_1 g_1 t_1 + S_2 g_2 t_2 + \text{etc.}}{S_1 H_1 + S_2 H_2 + \text{etc.}}$$

where:

S_1, S_2 are cross-sections of the prospecting holes;
g_1, g_2 are gravel thicknesses;
t_1, t_2 are gravel contents;
H_1, H_2 are thicknesses of gravel + non-productive material.

The weight of the useful substance contained in the deposits is thus equal to:

$$P = V \times Tm\ gr$$
$$P = V_1 \times Tm\ e$$

where:

P is the weight;
V is the total gravel volume;
V_1 is the total gravel volume + total non-productive volume;
Tm gr is the mean gravel content;
Tm e is the mean content of excavated material.

Despite their imprecision, these calculations allow the company some idea of the scale of the prospected deposit and over some guidelines for future exploitation.

3.3.1 ZONE OF INFLUENCE OF HOLES AND ROWS

The zone of influence of a prospecting row extends to the mid- distance of the two neighbouring prospecting rows located immediately upstream and downstream. A sample taken within a zone of influence thus defined is representative of the deposit for the area surrounding the sampling point.

3.3.2 ASSESSMENT

The skeleton triangulation method used to evaluate the contents is used for this purpose, possibly with a higher pit density. As a rule (the case of gold in particular), the distance between the rows must be 5 to 10 times the distance between the prospecting holes. If the valleys are narrow (less than 30m), then the spacing of the rows must be 25m and the pit interval 5m. For flats ranging in width from 30 to 100m the rows are spaced 50m apart and the holes 10m apart. If the prospecting holes indicate that the useful material is well-distributed throughout the gravel, the spacing between rows may be raised to 100m, but the spacing between holes remains the same.

If the contents vary widely from row to row and approach the workability cut-off level, 50m must be considered as a maximum. It is even advisable to reduce it to 25m with the holes every 5m if the river (or flat) has a heterogeneous nugget content.

The assessment group comprises 70 to 80 men. The assessment map (one per deposit) must be drawn to a scale of 1:1000. The maps should show the location of rocky outcrops, falls and rapids, the river and if possible its discharge, and for each hole, the thickness of non-productive material, gravel, gravel grade and grades in excavated material.

3.3.2.1 Calculation methods

Four methods are employed: the rectangle method, the trapezium method, the zone method and the isocontent curve method.

The rectangle method

This method **(Fig. 3.10)** yields results that approach reality. Subsequent control of operations is easy and rapid. In accordance with the principle of zones of influence

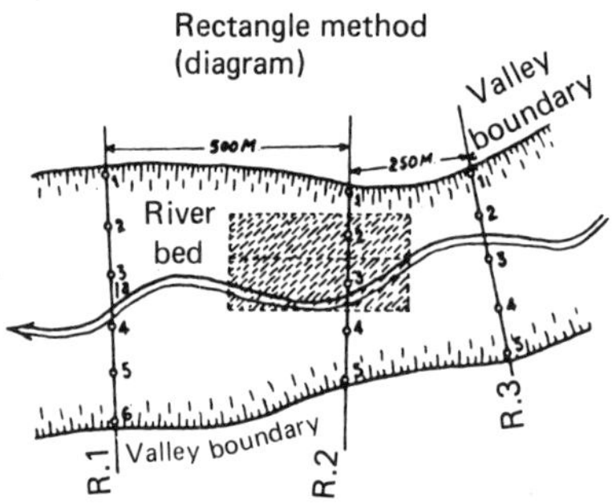

Fig. 3.10 *The rectangle method (from R Antione 'Les méthodes pratiques d'évaluation des gîtes secondaires: aurifères appliquées dans la région de Kilo-Mota').*

the mineralized zone is delimited on the map. The workable parts of the deposit are ultimately shown as one or more rectangles including all the works with a commercial interest. For a pit yielding approximately the limit content the boundary passes through the pit itself. For two neighbouring pits, one of which is commercial and the other is not, the boundaries passes through the middle of the space between them. At the end of the flat the boundary is equidistant between the first non-commercial row and the last commercial row and the boundaries are plotted perpendicular to the last commercial pit row. If a group of low-grade pits is found in a pay zone, it is always wise to surround it by a boundary. During the delimitation of the workable area it is recommended that a safety factor of less than 1 is used to account for recovery problems. This factor may vary from 0.64 to 0.90 depending on each specific case.

Similarly, in the case of isolated grades or isolated groups of commercial grades which are rare in the midst of non- commercial grades, it seems preferable to refrain from evaluating the assessment of the gravel and the corresponding quantities of useful substances because they are generally low and barely offer any hope of workability. In the case of isolated, exceptionally high grades (eg a nugget) located near normal commercial grades these contents must be reduced to avoid distorting the results (see the example of calculation using the zone method, the empirical reduction formula). In estimating the assessment large blocks are included in the gravel. It is necessary to apply a reduction factor related to the proportion of large blocks in the gravel volume remaining after the separation of these blocks.

Example

Consider three prospecting rows R1, R2 and R3 in a pay zone.
R3 and R2 are separated by a distance of 250m.
R1 and R2 are separated by a distance of 500m.

The pits are spaced at 10m intervals.

The prospecting log gives the following data:

Table 3.2. *Prospecting log*

Row No.	Hole No.	Grade g/m³ excavated	Gravel thickness	Non-productive thickness	Observations
R1	1				
	2				
	3				
	4				
	5				
R2	1	0.11	0.30	0.50	Distance between R2 and R1, 500 m
	2	0.50	0.50	0.30	
	3	0.65	0.60	0.50	
	4	0.03	0.30	0.80	
	5	0.03	1.00	0.70	
R3	1				Distance between R3 and R2, 250 m
	2				
	3				
	4				
	5				

The zone of holes 2 and 3 and row R2 is represented in the sketch (no scale) by the cross-hatched area. For each pit this provides a rectangle measuring 250m + 125m, hence 375m long and 5m + 5m, or 10m wide. The cut-off value is 0.2 g/m³ of excavated material (gold).

Thus the pay zone on row R2 is defined as follows:

Pit 1: Non-commercial grade No calculation required

Pit 2: Area 375m × 10m = 3750m²

Gravel assessment 3750m² × 0.50m = 1875m³

Non-productive assessment 3750m² ×0.30m = 1125m³

Total assessment 1875 + 1125m^3 = 3000m^3

Total gold 0.50g × 3000 = 1500g

Pit 3: Area 375 × 10m = 3750m^2

Gravel assessment 3750m^2 × 0.60m = 2250m^3

Non-productive assessment 3750m^2 × 0.50m = 1875m^3

Total assessment 2250 + 1875m^3 = 4125m^3

Total gold 0.65g × 4125 = 681g

Pit 4: Non-commercial grade No calculation required

Pit 5: Non-commercial grade No calculation required

Row 2: Total gold 1500 + 2681g = 4181g

Total assessment 3000 + 4125m^3 = 7125m^3

Mean grade 4181/7125 = 0.58g/m^3

excavated material

The same procedure is followed for each of the rows of pits containing commercial grades. The totals of all the commercial pits represent the reserves of the deposit concerned.

The trapezium method
After plotting the non-productive thickness, gravel thickness, mean gravel grade and mean excavated material grade on the assessment map for each of the holes all the holes located at the ends of the rows on the same side of the river are joined by a broken line. If the rows of holes are parallel, the areas thus bounded between two rows are generally trapezia **(Fig. 3.11).** In the curves the rows are no longer parallel. In this case the quadrilateral obtained is broken down into a trapezium and a triangle. The parallel sides of the trapezia thus remain oriented along the valley direction. The mean grade of the areas thus delimited is obtained by calculating the mean of the commercial grades of the area, with the cut-off value known (such as 0.5g gold/m^3).

The mean thickness of gravel and non-productive material is obtained by calculating the arithmetic mean of the thicknesses of all the holes in the trapezium, including non-commercial holes. The thicknesses thus obtained, multiplied by the area of the trapezium, give the total assessment of gravel and non-productive material for the area considered.

However, since a number of pits are non-productive, total assessment must be reduced by the assessment corresponding to the non-commercial holes. By applying the principle of zones of influence (for a hole in the middle of the flat the zone of

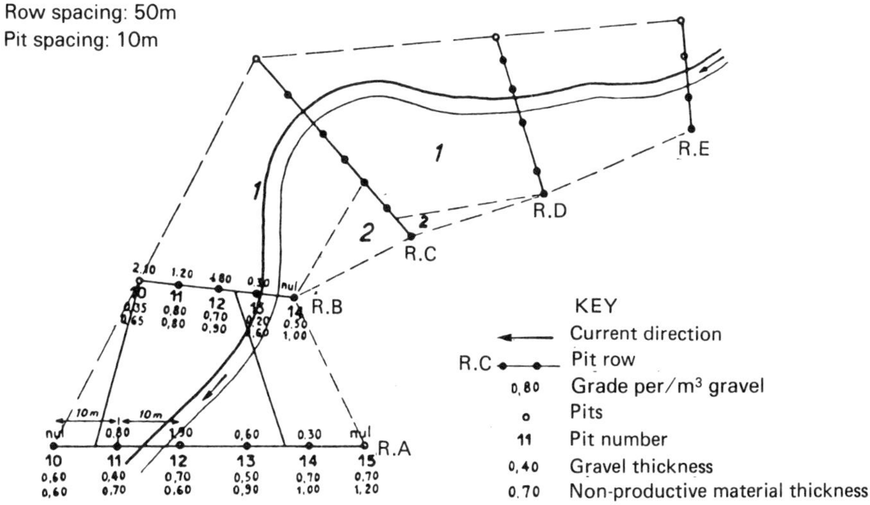

Fig. 3.11. *The trapezium method (from R Bonte 'L'étude des gîtes secondaires: métaux et pierres précieuses. Prospection et exploitation des alluvions').*

influence is a rectangle whose base is the distance between the holes of the row, and whose height is half of the distance between the rows surrounding it; for a hole at the edge of the flat the zone of influence is equivalent to that of a rectangle whose base is half of the distance between the holes of the row and whose height is half the distance between the rows surrounding it), the non-commercial areas are calculated first. The commercial area is obtained by subtracting them from the total area of the trapezium. The multiplication of this figure by the mean thicknesses of gravel and non-productive material yield the assessment of gravel and non-productive material in the workable zone. Multiplication by the mean grade gives the quantity of valuable mineral (gold, cassiterite, etc.) contained in the pay zone.

Example

Estimation of reserves lying between rows A and B, with the grade set at 0.5g/gold/m^3 gravel:

Mean content: sum of the products (gravel grade multiplied by gravel thickness) of the commercial pits in rows A and B, divided by the sum of the gravel thicknesses of the same pits:

$$\frac{(0.8 \times 0.4) + (1.9 \times 0.7) + (0.6 \times 0.5) + (2.1 \times 0.35) + (1.2 \times 0.8) + (4.8 \times 0.7)}{0.4 + 0.7 + 0.5 + 0.35 + 0.8 + 0.7}$$
$$= 2.02\text{g/m}^3 \text{ gravel}$$

Mean gravel thickness: sum of the gravel thicknesses in the commercial pits in rows A and B divided by the number of corresponding pits:

$$\frac{0.4 + 0.7 + 0.5 + 0.35 + 0.8 + 0.7}{6} = 0.57\text{m}$$

Mean thickness of non-productive material: sum of the non- productive thicknesses in the commercial pits in rows A and B divided by the number of corresponding pits:

$$\frac{0.7 + 0.6 + 0.9 + 0.65 + 0.8 + 0.9}{6} = 0.75\text{m}$$

Total area of trapezium: formula of the area of a trapezium whose base consists of rows A and B:

$$\frac{(50\ 40)}{2} \times 50 = 2250\text{m}^2$$

Area of non-productive pits: rectangular area determined from the zone of influence of all the non-commercial pits of rows A and B:

$$(5 + 10 + 5 + 10 + 5) \times 25 = 875\text{m}^2$$

Area of pay zone: total area of the trapezium less that of the non-productive zone:

$$2250 - 875 = 1375\text{m}^2$$

Volume of commercial gravel: area of the workable zone multiplied by the mean gravel thickness in the commercial pits:

$$1375 \times 0.57 = 784\text{m}^3$$

Volume of non-productive material in the workable zone: area of the workable zone multiplied by the mean thickness of the non- productive material in the commercial pits:

$$1375 \times 0.76 = 1045\text{m}^3$$

Gold reserve in the workable part: volume of commercial gravel multiplied by the mean grade of all the commercial pits in rows A and B:

$$784 \times 2.03 = 1591\text{g}$$

Mean content of excavated material in the workable zone: weight of gold in the workable zone divided by the total volume of alluvium in the pay zone:

$$1591/(784 + 1045) = 0.86\text{g}$$

The same procedure is followed for subsequent rows (forming trapezia and small triangles). The results obtained for each row of the deposit are added together to obtain the total number of kg in the prospected deposit.

The modified trapezium method

To take better account of circumstances liable to occur during mining the workable area is sometimes delimited to include isolated non-commercial pits located between commercial pits or at the periphery. This gives an area forming a deposit with a rationally workable content.

The zone method

The delimitation process is the same as in the modified trapezium method **(Fig. 3.12).**

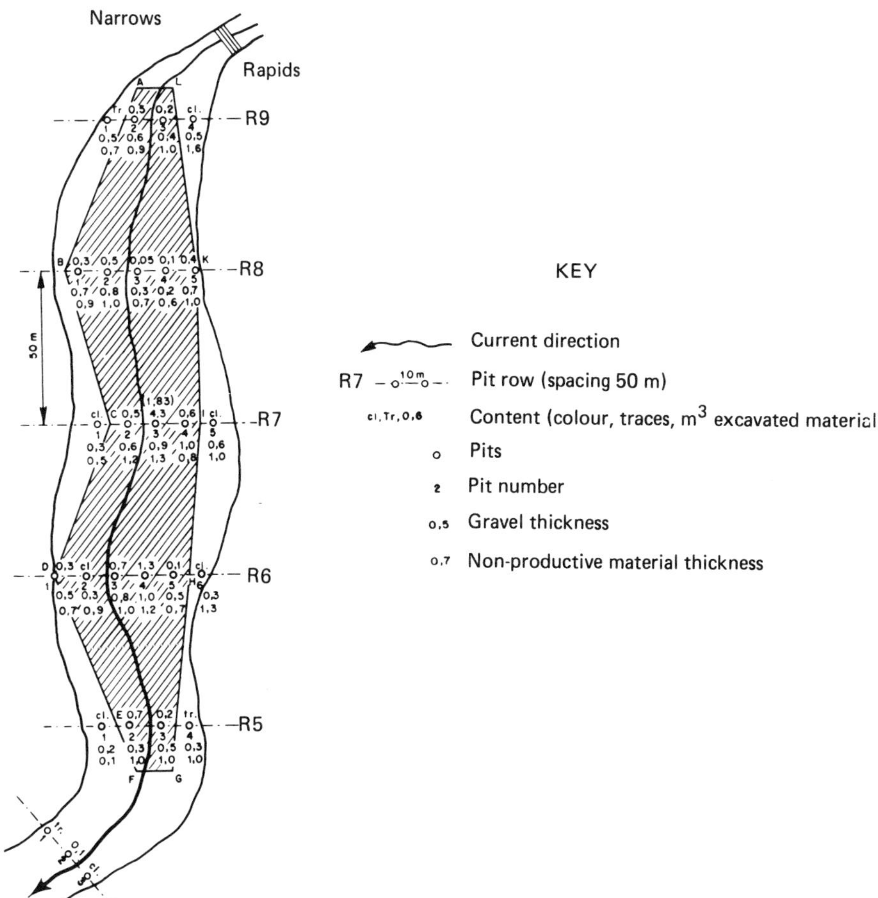

Fig. 3.12. *The zone method.*

Example

The flat extends between rows R4 and R9, considered as a zone (ABCDEFGHIKL). Rows are spaced 50m apart and pits 10m apart. Cut-off grade is set at 0.3g gold/m^3 of excavated material.

Row R9: the boundary of the zone passes half-way between pit 1 and 2 and half-way between works 3 and 4.

Row R8: the boundary of the zone is half-way between pit 1 and the edge of the flat half-way between pit 5 and the other edge.

Note that while they are not commercial, pits 3 and 4 are included in the same zone, because the mean grade remains commercial; in fact, the mean grade of excavated material from the pits of R8 is:

$$\frac{0.3(0.7 + 0.9) + 0.5(0.8 + 1.0) + 0.5(0.3 = 0.7) + 0.1(0.2 + 0.6) + 0.4(0.7 + 1.0)}{1.6 + 1.8 + 1.0 + 0.8 + 1.7}$$

$= 0.14g/m^3$ excavated material

Row 07: the boundary lies half-way between pits 1 and 2 and pits 4 and 5. Since pit 3 displays an abnormal grade, the empirical reduction formula used by certain prospectors is applied:

$$T = (0.4 \times t) + 0.15\ (t_1 + t_2)$$

where:

T is the assumed grade;
0.4 is the coefficient;
0.15 is the coefficient;
t is the gravel grade found;
t_1 and t_2 are the gravel grades from neighbouring pits.

Hence for pit 3 the grade is:

$$(0.4 \times 4.3) + 0.15\ (0.5 + 0.6) = 1.76g/m^3$$

This content is recorded in brackets on the assessment map.

Row R6: the boundary passes from the edge of the flat on the right bank to half-way between pits 5 and 6 on the left bank.

Row R5: the boundary passes between pits 1 and 2 and between pits 3 and 4.

Prolongation of the zone towards R10 and R4: the boundaries take account of the shape of the deposit, and are selected 10m upstream and 15m downstream.

Calculated areas of the different trapezia: total equal to $7475m^2$.

Calculated thicknesses:

Total mean = 1.50m
Gravel mean = 0.60m
Non-productive mean = 0.90m

Basin: .... Mission or site: .... Month: ....
River: .... Prospector: .... Map reference no: ....
Affluent: .... Substance: ....

Designation of work		Thickness (in m)			Volume of gravel in place V (m^3)	Volume of fine gravel expanded: R (gauges)	Volume of fine gravel washed: r (gauges)	Weight gathered p (g)	Contents			Observations
Line no. and spacing:	Well no. and spacing:	Sterile S	Gravel g	Total: $X = g + S$					per m^3 of gravel	per m^2	per m^3 excavated	– nature of bed-rock, – bed rock not reached, etc... – possible sketch
									$t = \frac{p.R}{V.r}$	$tc = t.g$	$te = \frac{tc}{X}$	

Table 3.3. *Systematic prospecting table.*

At: ....
Date: ....
Signature: ....

Table 3.4. *Systematic alluvial prospecting. Evaluation prospecting.*

Basin : .... Mission : .... Coordinates : .... Month : ....
River : .... Site : .... Map : .... References : ....
Affluent : .... Substance : .... Prospector : ....

Name number of river	Number of row	Azimuth of row	Distance between rows	Number of working	Surface acted on		Thickness		Volume m3 in place			No. pans (08 l) washed	P (g) substance collected	T1 content in m³ gravel	T2 content in m²	T3 content in m³ excavated	Qty substance per working	Qty substance per row	% boulders	Qty substance per row after correction	Total weight of substance	Nature of sterile	Nature of gravel	Nature of bed-rock (reached or not?)	Composition of panning concentrate	Observations
					by the working	by the row	Sterile	Gravel	Sterile	Gravel	Excavated															
TOTAL OF PAGE TO BE CARRIED FORWARD.																										

Calculated grades:

Mean of excavated material = $0.6g/m^3$
Gravel mean = $0.60 \times (1.50 : 0.60) = 1.26g/m^3$

The final assessment figures for the zone concerned are therefore:

Gravel assessment:	7475×0.60	=	$4485m^3$
Non-productive assessment:	7475×0.90	=	$6727m^3$
Total assessment:	$7785 + 6762$	=	$11,212m^3$
Total gold:	$0.60 \times 11,212$	=	6727g or 6.727kg

As above, all the essential data are recorded on the assessment map, including area, gravel assessment, non-productive assessment, total assessment, gravel grade, excavated material grade and total gold. The details must appear in the evaluation tables **(Tables 3.3 and 3.4)**.

The isocontent curve method
The same procedure is followed as in plotting the isocontent curves in tactical geochemical prospecting (Chapter 4). This method **(Fig 3.13)** is not widely used.

4 Eluvial prospecting

On the whole, eluvial prospecting is identical to alluvial prospecting. However, the prospecting methods may vary with the type of eluvial field and with the objectives that are set, namely, discovery of a workable alluvial deposit; exploration of a primary mineralized deposit covered by eluvium (vein, stockwork, lode, body, etc.).

4.1 Methods

The first case involves a simple attempt to locate and assess an eluvial deposit. Prospecting is conducted by the same principles and methods used in conventional alluvial prospecting.

The pit network in a region with an indeterminate geological structure (veinstone formations passing through rocks in all directions) or with a perfectly level surface (laterite platforms covering hilltops), is generally on a square grid. Where the mineralization is thought to be in parallel bands, the square grid is replaced by a rectangular grid elongated in line with the assumed general trend of the deposit. The dimensions of the prospecting grid are selected in accordance with the area to be prospected. The pits are spaced at intervals of 20 to 30m or at a maximum of 100m. The grid is narrowed if the mineralization proves to be highly irregular. If the contents display little variation, this narrowing is only carried out at the assessment stage.

A primary deposit is located in accordance with:

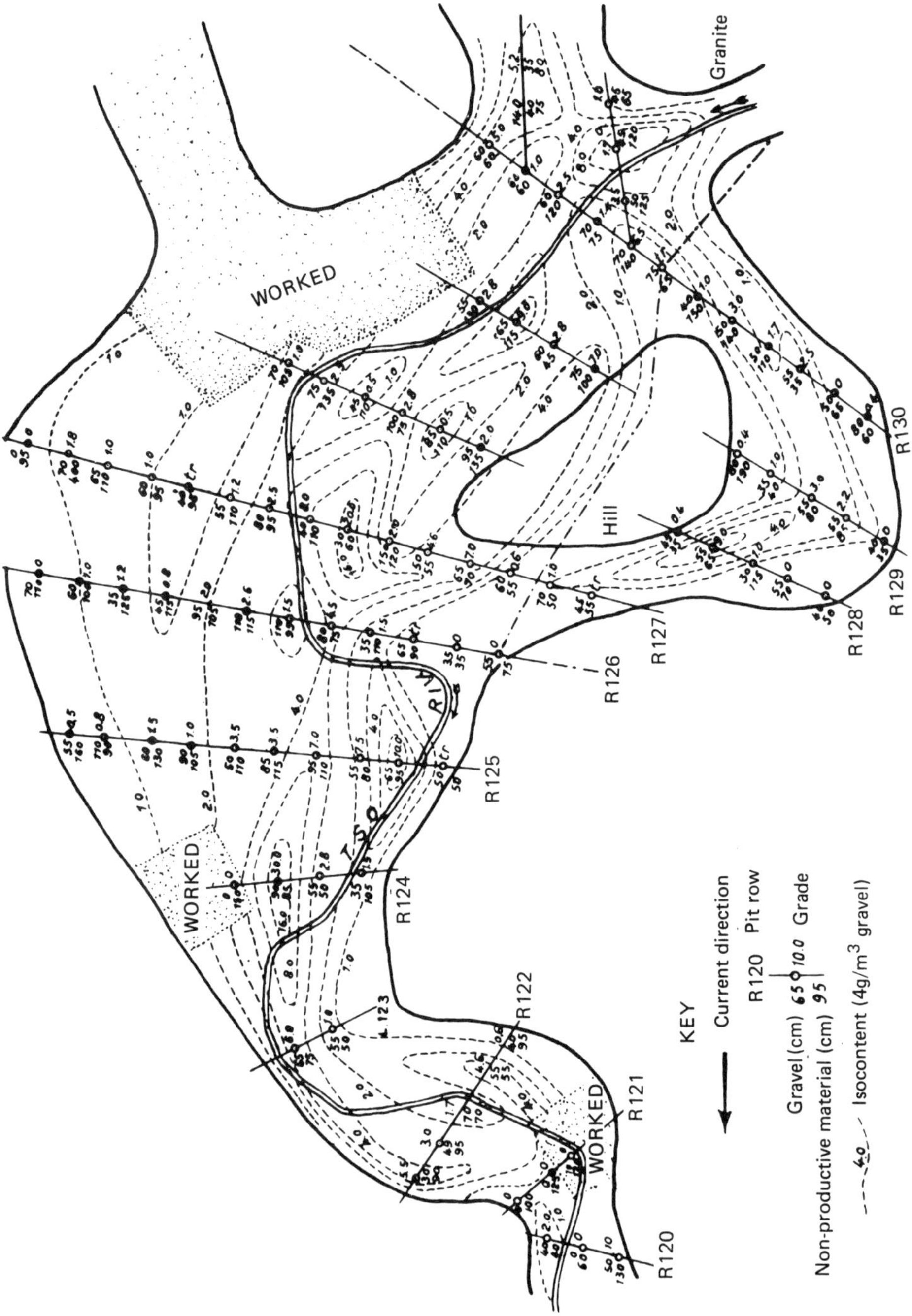

Fig. 3.13. *Example of a plot of isocontent curves (from R Antione 'Les méthodes pratiques d'évaluation des gîtes secondaires aurifères appliquées dans la région de Kilo-Mota').*

1) the increasing range of eluvial grades;
2) the sharpness of the outlines or angular forms of ore grains (or accompanying minerals);
3) the growing size and increasingly angular appearance of the rock fragments forming the gangue of the primary ore;
4) their concentration.

In a region with a moderate relief and no forest cover a preliminary photogeological survey can be used for the location of the prospecting sectors. On the other hand, in a zone with dense forest cover the primary deposit is approached by locating the zones of ore concentration and fragments or blocks of angular gangue, and then narrowing the distance between the pits, insofar as no gossan or outcrop is visually detectable.

The discovery of the primary occurrence marks the final point of eluvial prospecting. The deposit found then gives rise to another exploration phase executed by different methods, applicable to the investigation of rock deposits (veins, stockwork, lodes, etc.) (Chapter 2, Chapter 7.)

The approach to a primary deposit is much more easily done by geochemistry and is generally done by this method (Chapter 4).

4.1.1 HELPING TO LOCATE A PRIMARY DEPOSIT

The sudden disappearence of mineralization and of angular blocks of gangue on a moderate slope covered by recent eluvium indicates a primary deposit.

The presence of mineralization on both sides of a ridge indicates that the primary deposit occupies or intersects the ridge.

In an old eluvial deposit (reworked laterite), forming a shield over a completely flat area, only the commercial parts of the eluvial deposit are assessed. The primary deposits are generally discovered during the mining of the deposit.

4.1.2 ELUVIAL MINERALIZATION

Eluvial mineralization can present in the following ways: suspended in eluvium as the result of the disintegration of the host rock and movement over a short distance (deposit); as inclusions in gangue fragments from the primary deposit, where the eluvial soil is non-productive (scree, rock debris) or as a combination of these two states (mixed deposits). In the case of a mixed eluvial deposit the free useful grades of the gravel and the mineralized fragments are determined. The sum of these two figures represents the total grade. If mineralization is only apparent in rock fragments, analyses are performed similar to the evaluation of primary deposits.

4.1.3 DIGGING PITS AND SURVEILLANCE

During pit digging operations the cuttings are separated from the different layers of the eluvium, and if no bed is visible, the cuttings removed are separated metre by metre (50cm by 50cm if necessary). Since each pit is perpendicular to the line of greatest slope, the first pile of cuttings is placed uphill and the others in order downhill from the pit. As a rule, the pits are stopped as soon as the bedrock is reached, after which 15 to 20cm are extracted.

In general, all the gravel extracted from each pit is washed in a sluice or jig, after the mud has been removed. If mineralization is free in the eluvial soil, it is recovered directly from the sluice bottom. The clean gravel remaining in the sizing tank or in the mud removal tank, and the gravel recovered at the outlet of the sluice, are examined carefully and stored separately for subsequent analyses. If mineralization is composite (free grains of mineral and mineralization in rock fragments), part of the total grade is determined by sluicing.

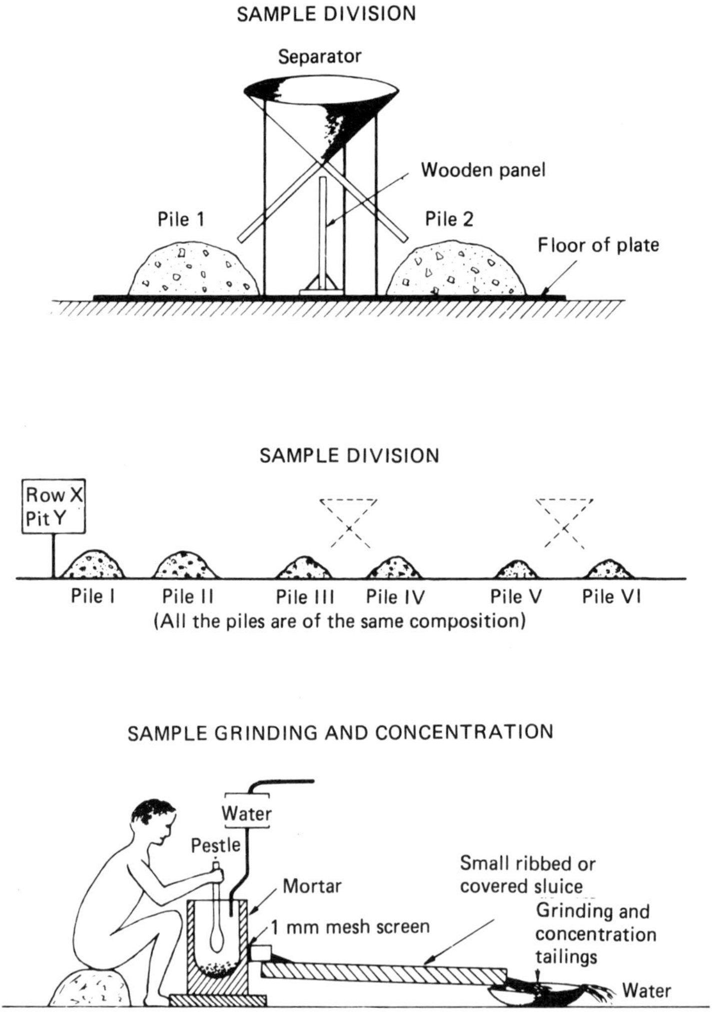

Fig. 3.14. *Sample preparation (from P Tkatchenko 'Instructions génerales pour la prospection éluvionnaire et filonienne').*

The remainder, after storage, assessment and sampling of mineralized rock fragments, is obtained by grinding and concentration. If the mineralization is only included in rock elements, washing the eluvium eliminates the soil and separates the mineralized rock fragments on the screens. They are then assessed and sampled. This process is used to sample old dumps.

4.1.4 GRINDING

Even in the case of free mineralization in eluvial soil, it is advisable to test the rock elements of the eluvium by grinding and panning. With respect to gold, the superficial laterite sometimes contains free gold, which can only be detected by grinding to a fine powder and thorough panning. Soft rock elements mixed with the eluvium are also liable to contain mineralization (decomposed pyritized shales, bedrock cut by a stockwork or by quartz or pegmatite lenses, etc.).

The following procedure must be used to prepare an analytical sample **(Fig. 3.14)**. The oversize from the perforated plate of the mud removal pit and the tailings collected at the outlet of the washing plant are broken by hammer to 50mm pieces, implying that a maximum of 5 to 10% of the total weight of the batch to be sampled is retained by a screen with a mesh size of 50mm. They are collected together on a wooden floor or clean metal plate and mixed using a shovel to form a single homogeneous pile. Using a large 'divider', the pile is divided into two approximately equal piles. Using the same separator or a smaller separator one of the piles obtained is again divided into two. This division is continued until two piles are obtained of about 100kg each (about 15 8 litre pans of expanded gravel). One of the last two piles thus prepared is dried and used to prepare several concentrated homogeneous samples for analysis. The other piles are used as controls. They are labelled and set aside for check tests. After hammer grinding on a thick plate or in a mortar to 5mm pieces (mesh size of the screen retaining 5 to 10% of the pile to be ground) all the pieces of the sample are mixed thoroughly on a sheet of tarpaulin by raising opposite corners of the sheet, in succession. The pile of crushed ore is then divided by a smaller separator until two 3kg samples are obtained. One of these samples is sent for tests to determine grades. The second is stored in a numbered small sealed crate.

Carried out this way, according to the nomograph of P Gy, this sampling method offers data containing a relative error of 100% for gold contents, which is quite satisfactory if the grade found is less than half or greater than double the grade set in advance as the limiting grade.

4.1.5 SAMPLE CONCENTRATION

The sample is ground to 1mm in a light-weight grinder (35kg total weight) used for rapid prospecting. In the case of wet grinding, the product exits automatically and is concentrated as grinding proceeds, in a small sluice. When the entire sample is ground, the useful substance is collected (clean-up). The mortar is first turned upside down and all the heavy concentrate is collected, while washing the screen and pipe thoroughly. The corduroy strips or the lid are then removed and washed in the pan which already contains the heavy mortar concentrate. The small sluice is then cleaned and all the concentrate collected in the same pan. Dry grinding necessitates frequent screening of the product to obtain the desired particle size. Cleaning of the grinder raises problems in this case and must be monitored constantly. Concentration only takes place in a pan.

The concentrate is reconcentrated once in a small pan (or buddle), and then dried, cleaned with a magnet and blown. If the concentrate contains amalgable gold, the latter is recovered before drying by adding mercury directly to the small pan, if some is available. If the gold-bearing concentrates are bulky, then they are poured into a clean bottle with about half a litre of water and a drop of mercury. The bottle is closed and shaken for 30 minutes. The amalgam is extracted from the mass of concentrate and wrapped in a sheet of paper. The concentrate remaining after amalgamation is mixed with the grinding products, to serve for the laboratory determination of the combined gold grade by lead fusion. The amalgam is heated in a special graphite-lined retort to prevent the gold from sticking to the walls. The mercury is generally recovered in another retort after evaporation.

4.1.6 GRADE

The useful mineral content of the rock elements in eluvium is determined after direct weighing on a precision balance of the very clean concentrate using the following equation:

$$t = \frac{\text{Po} \times \text{ton (in g)}}{\text{Pe}}$$

where:

t is the grade in g
ton is 1,000,000g
Po is the weight of useful mineral collected in g
Pe is the weight of ground sample in g

The content is expressed in g per ton of raw ore.

4.1.7 ASSESSMENT OF AN ELUVIAL DEPOSIT

Depending on the type of eluvial deposit the reserves are evaluated on the basis of different principles:

1) CASE OF A DEPOSIT in which mineralization is included exclusively in the eluvial soil in the form of free mineral grains: the calculation methods are identical to those used for alluvial prospecting.
2) CASE OF DEBRIS OR SCREE in which mineralization exists only as inclusions in rock elements of the eluvium: the calculation is based on the estimation of the total commercial tonnage of mineralized rock elements which, multiplied by the grade per ton, gives the total weight of valuable mineral.
3) CASE OF MIXED DEPOSIT in which mineralization exists in the soil and rock elements: an estimate of the reserves contained in the soil is established

in accordance with the principles adopted in alluvial prospecting, followed by a calculation of the weight of valuable mineral in the rock (similar to the previous case).

5 The specific case of diamond prospecting

Alluvial prospecting may be considered as a direct or indirect method to discover diamond-bearing deposits. In the former case, diamond is searched for directly in the alluvium, where the objective is the detection of an alluvial deposit. In the second case, an attempt is made to find one or more minerals which may lead to the location of the original host rock of diamond, kimberlite.

5.1 Prospecting for alluvial deposits

In this case, prospecting is conducted by the techniques discussed above, by tests in the active stream bed and/or by pits in the bedrock. However, a number of specific points are applicable.

1) The choice of gravels to be washed is decisive in strategic prospecting phases: as a rule, the more a gravel is reworked, the more it is reconcentrated and the easier the detection of diamond. Hence the active stream bed gravel is preferable to gravel taken from below the banks which is itself preferable to terrace gravel. Note that the active stream bed gravel is also:
   easier to sample (no overburden);
   easier to wash (contains very little clay in comparison with bank or terrace gravel).
2) Volumes of gravel to be washed are often very large because the prospected grades are low: a grade of 0.50ct/m^3, which already constitutes a reasonable diamond grade, corresponds to 0.05ppm, and also because only a limited number of stones are required to obtain these grades or even a fraction of a stone if the stones are large.

   Example

   Consider a gravel grade 0.2ct/m^3, in which the stones have an average weight of 0.50ct, giving two stones for one carat. Statistically, to find one stone it is necessary to treat 2.5m^3 of gravel.

   Prospecting samples of 1m^3 would have:

   67.05% chance of finding 0 stone;
   26.81% chance of finding 1 stone;
   5.36% chance of finding 2 stones;
   0.93% chance of finding 3 stones.

It is obvious that the smaller the number of stones, the easier they are to detect: hence for gravel containing 0.2ct/m^3, if the average weight is only 0.20ct, no more than 1m^3, is necessary to find one stone.

3) Security is one of the fundamental concerns in diamond prospecting. The prospector must be constantly aware of the many possibilities of theft. This is certainly the greatest problem in in this case, which combines simple light-weight equipment with maximum security.
4) Gravel treatment techniques involve two types of operation: classification and concentration.

5.1.1 CLASSIFICATION

This is carried out by reducing the meshes (eg 10, 5, 2.5, 1mm) using screens. The equipment used varies from being simple to highly sophisticated: round sieves, rocking screens, tables, sizing drums, vibrating screens. The last two are motor driven.

5.1.1.1 Standard procedure

The gravel is extracted and classified using a series of round sieves. The screeners are placed in the water, in a row, downstream to upstream, so that the bowls can be returned by simply floating them along the water **(Fig. 3.15)**. The operation begins

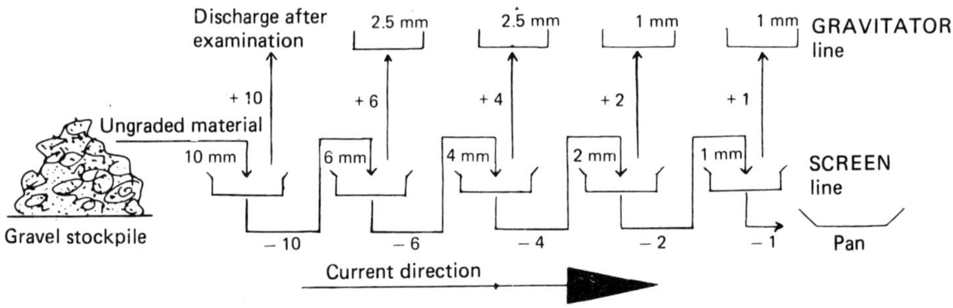

Fig. 3.15. *Example of a gravel treatment flow sheet using a screen and a gravitator.*

by removing the stones on a 10mm screen by the first man in the row (upstream). The oversize is dumped after examination. The undersize is collected in a bowl under water. The first screener then sends it to his neighbour who retrieves the sieved material from a 5mm screen, etc. The oversized pieces too big to pass through a particular screen are stored on the bank near the particular screener (in a bowl or on a tarpaulin) and then concentrated.

5.1.2 CONCENTRATION OF MATERIALS

Concentration is carried out using a gravitator **(Fig. 3.16)**, the jopling-jig **(Fig. 3.17)**, the BRGM hand jig, the Plietz jig, the BRGM motor-driven jig (diameter 315) or any other type of jig. The important factor is not the type of instrument but the way in which it is used. In all cases, the classification is densimetric.

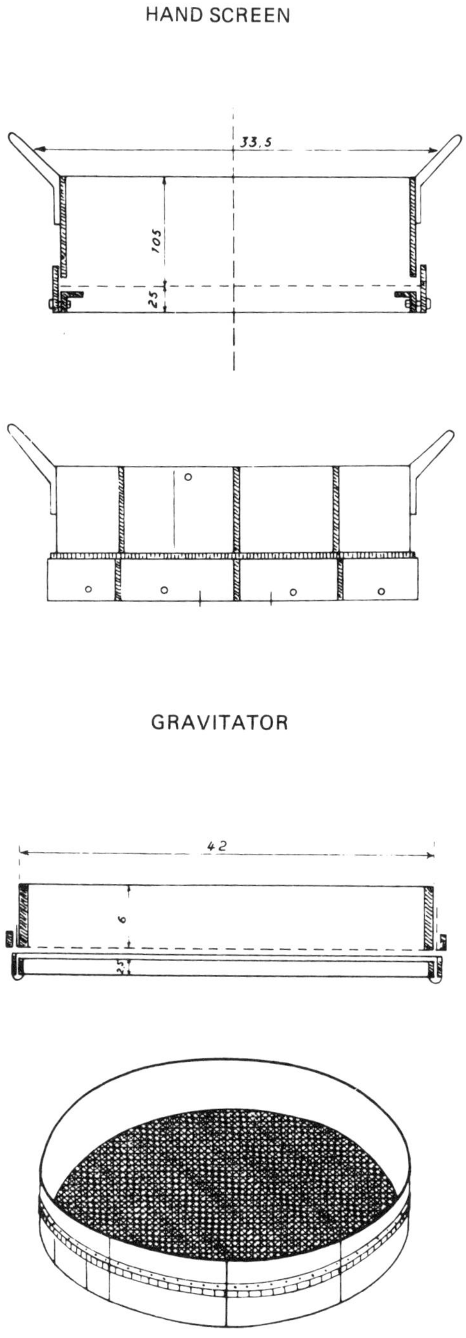

Fig. 3.16a. *Gravitator.*

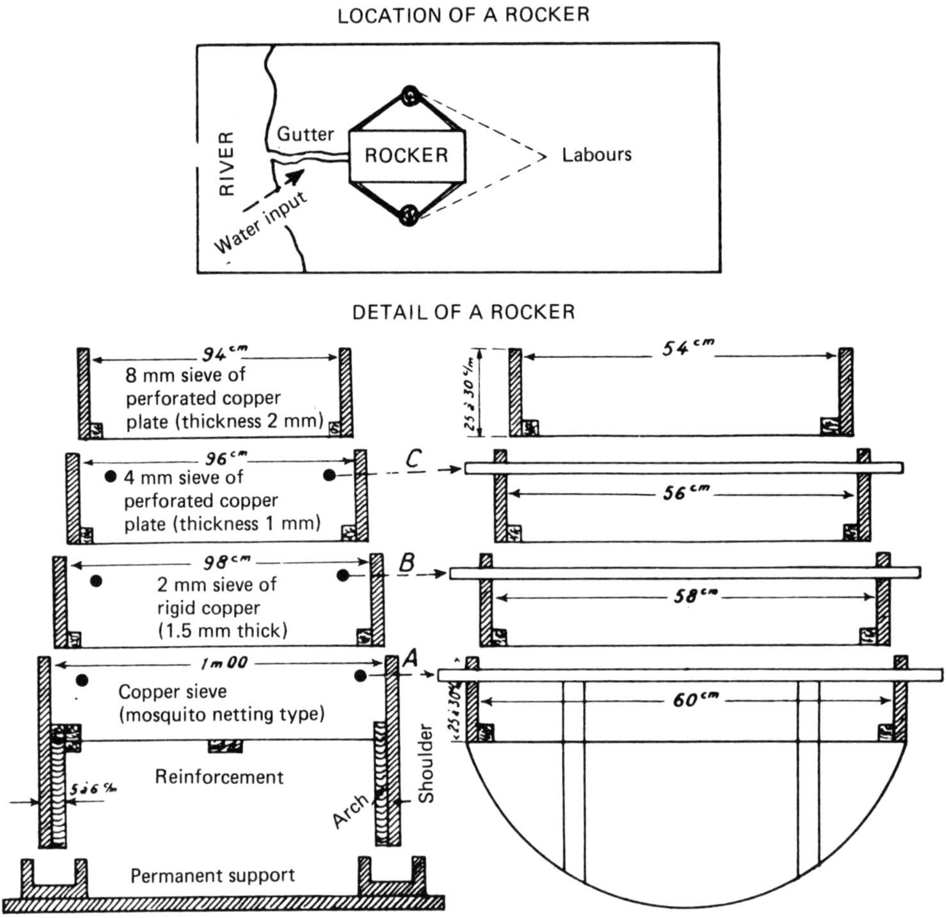

Fig. 316b. *Rocker (from J Goor 'Prospection des gemmes et diamants').*

5.2.1.1 Standard gravitator procedure (or the Brazilian peneira)

Each category of oversized pieces is concentrated in gravitators with metal screens of corresponding mesh (and also circular, light-weight screens made of wood or aluminium), kept directly vertical at a certain depth of water. The operation begins by giving the gravitator a few small vertical jerks and then actuating it with a rapid descending motion together with rotary movement through an angle of about 20°. Everything takes place as if the gravitator was suddenly immersed in the water. It is then raised slowly to its initial position. The operation is then repeated several times. The heavy elements gravitate to the bottom and form a sort of ring, while the heavier elements are concentrated at the centre.

Once concentration is completed, the gravitator is overturned in one stroke on to a flat surface. If the operation is performed correctly, the heavy minerals are found at the top of the cake thus formed in the centre, in a ring a few cm wide. The

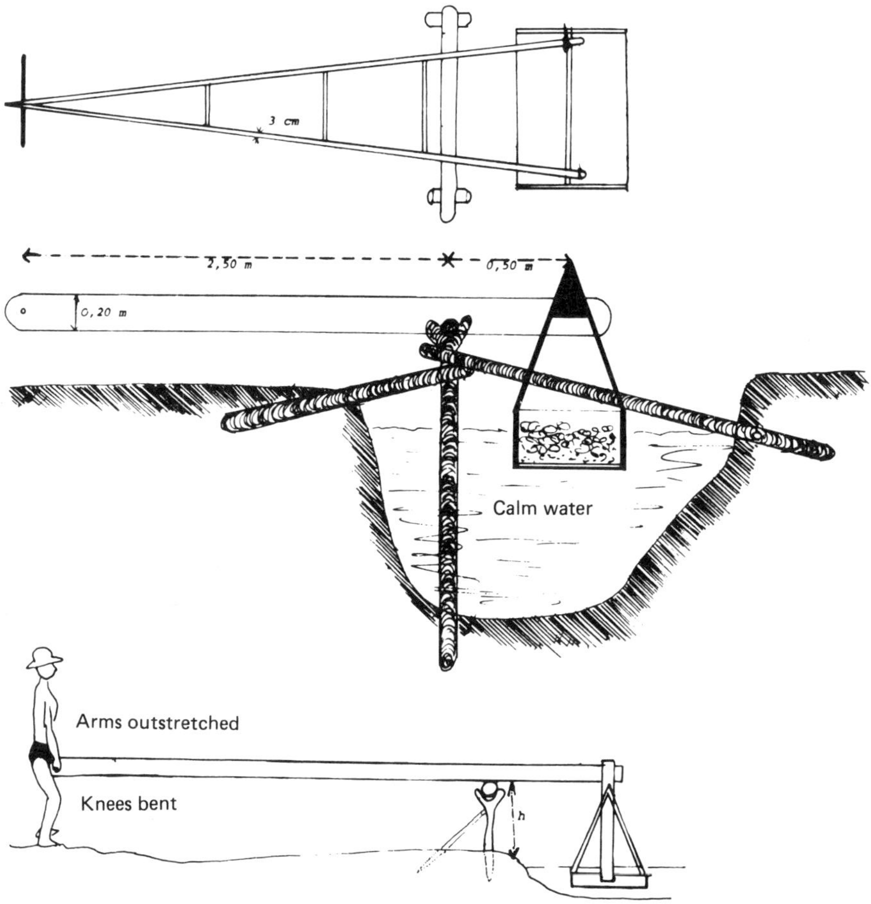

Fig. 3.17. *Prospecting jig.*

size of the ring (diameter in cm) varies with the proportion of heavy phase in the gravel concerned.

5.1.3 EXAMINATION OF CONCENTRATES

In this stage, it is important to examine the heavy minerals carefully because although unrelated evolutionary they are a product of the geological environment and some of them may be accompanied by diamond.

The minerals in the cake are examined, from the periphery towards the centre. Interesting grains are sampled, particularly the central disc. Obviously, it is at the centre of this disc that diamond must be sought, but the concentrates are examined in the shade, under 1cm of water, in a basin with a black bottom. Diamonds give a characteristic bright shine, whereas other minerals which shine in the sunlight (zircon, chrysoberyl, quartz) are dull under water. The results are recorded on a test

Table 3.5. *Test form.*

Example

Test N° ☐
General N° ☐

Territory	Region	Basin	Map ref.

Mission	Prospector	Report

Date ☐

ACCESS

Sub-affluent : ....
Affluent : ....
River : ....
Left – Right – bank – Buried Channel

Surface : ....m^2
Sterile thickness : ....m
Sterile Vol. : ....m^3
Gravel thickness : m
Gravel Vol. : ....m^3

Type of test	Method of washing
Damming	Pan
Potholes	Sieve
Buried channel	Gravitator
Sub-bank wells	Rocker
Flat wells	Hand jig
Terrace wells	Motor jig
Slope wells	

Distance to river ....m

River : Width ....m
Depth ....m
Current ....

Flat
Width : Right ....m
Left ....m

Occurrence of water : YES – NO
Timbering : YES – NO

Total depth : ....m

Expanded cubage washed in litres ☐

Granulometry	Roundness	Type	Nature	Cement
Verse coarse	Well-rounded	Sandy	Quartzose	Laterilized
Course	Rounded	Sandy shale	Polygenetic	Siliceous
Medium	Sub-rounded	Silty	Polymietite	
Fine	Sub-angular	Clayey	Lateritic	
	Angular		Schistose	

Bed-Rock Reached : YES – NO – PARTLY

Nature : .... Quality of stripping : ....
Direction of dip : .... Surface stripped : ....
Degree of alteration : .... Depth stripped : ....

Sample B.R.

Concentrated Abundance:

Minerals : + 10 ....
– 10 + 5 ....
– 5 + 2.5 ....
– 2.5 + 1 ....
– 1 ....

Diamond

Stones : ....
Carats : ....
S/ct : ....
C'/m^3 gravel : ....
Stones > 1 ct
Number : ....
Weight : ....

Characteristics of gravel components

Examination

Drilling : YES – NO
Washing : YES – NO

Value of test

Very good
Good
Average
Bad

Table 3.6. *Form for recording state of monthly progress.*

EXAMPLE

STATE OF MONTHLY PROGRESS

Prospector__________ Month of__________ 19____

Sector	Test No.	Surface	THICKNESS			VOLUME			STONES 1 ct		STONES 1 ct		TOTAL DIAMONDS		WEIGHT IN CARATS (1/5 g)	CONTENT Ct/M^3 Gravel
			Sterile	Gravel	Treated	Sterile	Gravel	Treated	Number	Weight	Number	Weight	Number	Weight		

form **(Table 3.5)** and summarized in a monthly progress report of tests and pits **(Table 3.6)**.

5.2 Prospecting for kimberlite deposits

Alluvial prospecting can be used as an indirect method to identify the location of primary diamond-bearing deposits, namely kimberlite.

The alluvial concentrates are not only analysed to find diamond, but also the so-called indication mineral deposits which help to find the host rock. The most frequently used mineral is magnesian ilmenite or picroilmenite. Chromites rich in chromium, pyrope garnet, olivine or chrome diopside are also used. These minerals, which are more fragile, are also difficult to use as tracers, because when they are present, they are only found near the primary occurrence. The prospector should be aware of these minerals before setting out.

In strategic alluvial prospecting, if the alluviation conditions are satisfactory, two or three pans of gravel (about 20 litres) are sampled, less than 5mm, with a density of 0.5 to 1 sample per km^2 in accordance with drainage. Fractions between 5 and 1mm are classified and concentrated on a screen. The heavy kimberlite minerals are separated and the grains counted in the coarse fraction. The fraction under 1mm is panned and sent to the laboratory.

In actual fact, alluvial prospecting is the first stage in finding kimberlites which must be supplemented by techniques offering better resolution power, such as soil samplings (or 'loaming'), first with an overall survey and then a systematic study. This can be combined with various geophysical methods and with the interpretation of aerial photographs.

5.3 Equipment for diamond prospecting

The equipment is often bulky depending on the type of prospecting and its objectives, access problems and transport facilities (see also Appendix 1 to Chapter 1):

1) buddles (8 litre and 2 litre pans);
2) screens;
3) gravitators, 'peneiras' (mesh size 1, 2, 4 or 1, 2.5, 5 and possibly 0.5mm);
4) metal brushes;
5) plastic bags 30 x 18cm;
6) plastic tanks;
7) bowls.

For kimberlite prospecting in the field:

1) metal scrapers (10 x 5cm);
2) canvas or strong plastic bags (to transport the samples to the water point);
3) small square screens for dry screening.

For mineral analysis:

1) sample tubes;
2) tweezers;
3) binocular microscope;
4) set of laboratory screens (diameter 10cm);
5) heavy liquids, filters, beakers;
6) 10kg suction magnets;
7) Petri dishes.

For pit sinking:

1) drill rod;
2) crowbar;
3) helmets;
4) ropes;
5) shovels, picks;
6) metal buckets;
7) pulleys (4m, 6m);
8) motor pump;
9) casing etc.

5.4 Grade evaluation

To evaluate the diamond content of a flat or a kimberlite it is nearly always necessary to treat 100 to 300m^3 of gravel. For this purpose, the BRGM has developed a mobile washing plant for mineral concentration using gravity, which can also be used for alluvial mineralization other than diamond, including cassiterite. Mounted on a trailer or on a barge as required it can include the following components:

1) bucket elevator;
2) chain mud cleaner;
3) selective roll grinder and rod grinder;

4) vibrating sreen with four sieves for sizing grains;
5) screening board to separate fines and eliminate wash waters;
6) series of four to six circular jigs with dense filter bed;
7) constant level extractor for automatic removal of overburden;
8) riffles box, sluice type, to work out the oversize;
9) electric motors or heat engines and various accessories.

This washing plant is designed for a throughput of 2 to 6m^3/h. The recovery of heavy minerals, including diamond, without human involvement, facilitates the use of security systems designed to prevent theft.

6 Prospecting coastal deposits

6.1 Phase one: rapid prospecting

6.1.1 INTRODUCTION

Rapid or summary prospecting is carried out at the start of exploration. This is known as phase one. Very rapid, and therefore economical, it draws on simple, manual, light-weight drilling equipment, generally consisting of Helix type augers or the Hand Auger Kit manufactured in Australia and designed for this type of formation. Its purpose is to cover the largest possible area in minimum time, to orient subsequent prospecting and to distinguish in a mineralized sandy area the enriched zones which should be tested in detail in the second phase. Since the drill holes are not cased, they generally cannot reach the bedrock and stop (in principle but not necessarily) at the water-table. This is a minor drawback in the preliminary phase because, as a rule, mineralization of coastal beach formations has already undergone a preclassification by wind action. Hence the grades taken are, in principle, higher in the surface levels than in the lower levels. This is often true in coastal formations, especially in Africa.

6.1.2 SAMPLING GRID

The sampling grid is generally about the following size:

1) row spacing : 1000 to 1500m;
2) pit spacing : 100 to 50m minimum.

This layout is imposed for the two following reasons:

1) As a rule, coastal formations consist of more or less parallel strips, often following a necklace shape and rarely wider than 100 or 200m.
2) For mechanized mining of these mineralized deposits to be profitable (especially using a dredge) the width must not be less than 70 to 80m.

The layout of the pits according to this grid, on rows perpendicular to the main axis of the mineralized deposits, always allows one pit to be placed near the centre of the mineralized deposit, followed by two pits on either side of the formation investigated. The information thus obtained is sufficient to orient rapid prospecting with maximum effectiveness, even if the bedrock has not been reached in the first phase.

6.1.3 SAMPLING METHOD

Sampling is carried out along the entire length of the drill hole, metre by metre (in case of strong mineralization), or by selecting sample lengths according to visual estimation of the degree of mineralization.

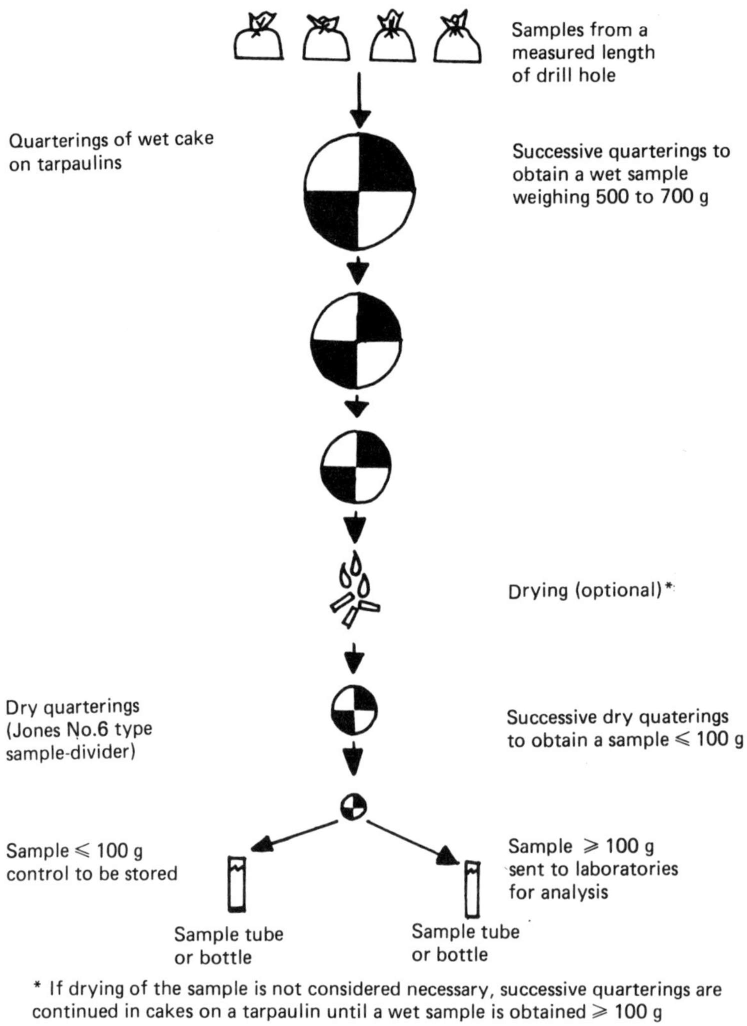

Fig. 3.18. *Direct despatch of analytical samples to the Laboratory. A measured length can be taken to be 1m throughout this book.*

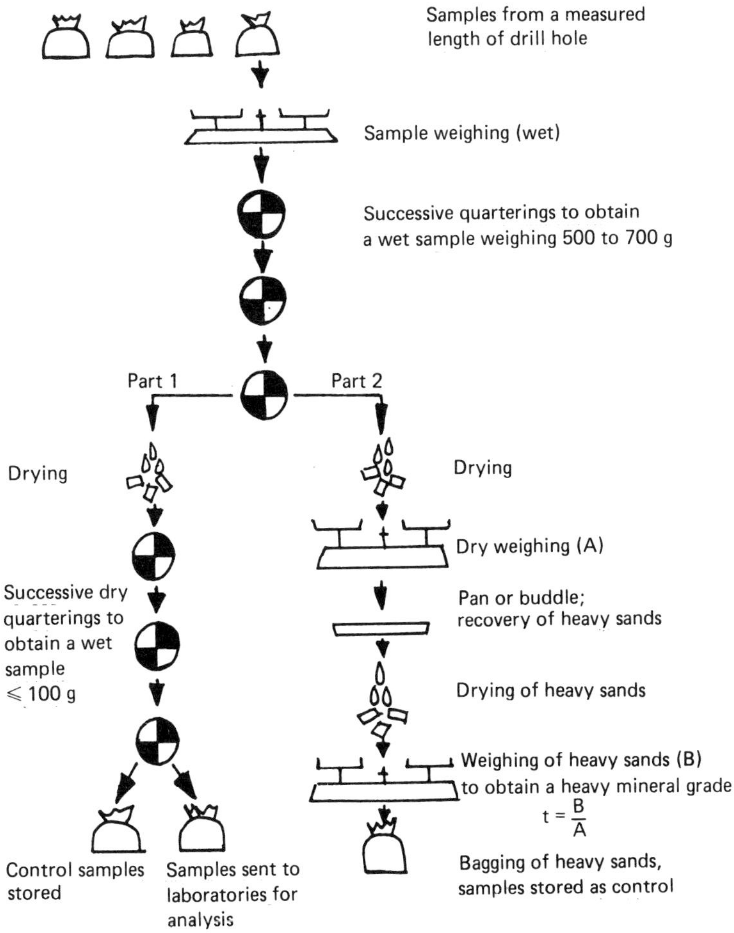

Fig. 3.19. *Pre-treatment of samples in the field for the immediate determination of heavy mineral grade.*

The samples obtained are reduced by successive quarterings to about 100g of ungraded ore (ie the contents of a tube 35cm in diameter and 7cm high) for laboratory analysis, to determine the percentage of heavy minerals contained in the ungraded ore (ie mineralized sand taken from the drill hole). The quatering of the drilling product can be carried out on the spot on small tarpaulins or at the base came.

6.1.3.1 First case

In the case of direct despatch of samples to the central laboratory for analysis **(Fig. 3.18)** the samples taken metre by metre on each drill hole with sorting of the enriched sections must be classified and if necessary dried and quartered until two samples of 100g each are obtained. One is sent for analysis to the central laboratory and the second stored as a control.

6.1.3.2 Second case

The following procedure can be used if the grade of the heavy mineral must be determined immediately. In case of pre-treatment of samples at the field laboratory **(Fig. 3.19)**, followed by the joint despatch of the same analytical samples to the central laboratory, the same procedure is followed until a wet sample weighing about 500 to 800g is obtained by successive quarterings. This sample is divided into two fractions, one of which is sent to the central laboratory for analysis and the second stored as a control. The second half, after drying and weighing, is treated by the buddle. The heavy sands are weighed, making it possible to obtain the contents and percentages on the spot, placed in bags and stored for the subsequent preparation of composite concentrates, if required. As an economy measure the laboratories are not usually sent samples that are insufficiently mineralized or below the cut-off grade, while remembering that a 5% mineralized sand appears non-productive to the naked eye! The distance drilled corresponding to the laboratory samples is recorded.

6.1.4 REQUEST FOR LABORATORY ANALYSIS

The following should be requested from the laboratories:

1) Heavy mineral content of the ungraded ore. This grade is expressed as a percentage by the ratio:

$$t = \frac{\text{weight of concentrate}}{\text{weight of ungraded ore}}$$

2) The mineralogical composition of the concentrate (rapid approximate mineralogical determination).

6.1.5 CONCLUSIONS DRAWN FROM RAPID PROSPECTING IN PHASE ONE

Phase one rapid prospecting procedures (presented above) make it possible to quickly and economically locate and circumscribe the most mineralized zones, as well as a priority choice for their study by semi-systematic prospecting. It thus helps to avoid wasting time on non-productive areas, where the heavy mineral grades are of lesser interest. This is especially important since exploration studies generally cover a very wide field area. It enables initial estimations of the size and weight of mineral deposits in the formation studied.

6.2 Phase two: semi-systematic prospecting

6.2.1 INTRODUCTION

After rapid prospecting has pinpointed the mineralized areas a more detailed study is made of the circumscribed deposits.

Drilling in coastal sandy formations must be conducted beyond the water-table down to the bedrock. Helix type drills or the Australian Hand Auger Kits are replaced by more complex equipment, which allows penetration through the shallow ground water using cased bore holes and light-weight Australian hand drills of the 'below water-table' type, for example.

6.2.2 SAMPLING GRID

Semi-systematic prospecting employs a tighter sampling grid:

1) In formations with high sand volumes and little mineralization (recent eolian or dune coastal formations):
   — row spacing: 640m, subsequently reducible to 320m;
   — pit spacing: 80m, subsequently reducible to 40m in the richest seams.
2) In well-mineralized formations (ancient coastal formations):
   — row spacing: 320m, subsequently reducible to 160m or 80m;
   — pit spacing: 40m, subsequently reducible to 20, 10 or even 5m in very rich seams.

The drill holes are placed in rows perpendicular to a baseline approximating to the main axis of the mineralized formations investigated.

6.2.3 SAMPLING METHOD

Sampling in semi-systematic prospecting is more complicated than in rapid prospecting and requires sustained supervision of the different treatment operations which must be conducted with meticulous care. Two operations are involved:

1) treatment of samples in the field to obtain the grade of the concentrate from each drill hole through each mineralized seam;
2) preparation of composite concentrates for analysis by the central laboratory.

6.2.4 SAMPLE TREATMENT AT THE FIELD LABORATORY

6.2.4.1 Heavy minerals grade (concentrate) determination

Sampling is initially carried out in the field laboratory (**Fig 3.20**). The mineralized sand sample is treated in the pan or buddle to obtain the grade of the concentrate in each measured length of the mineralized section of each of the drill holes. Throughout this book a measured length can be taken to be 1m. Occasionally, a small shaking table may be employed. The mineralized sand grade expressed as a percentage is:

$$t = \frac{\text{weight of concentrate obtained in the pan}}{\text{total weight of sample}}$$

Preparation for sampling in semi-systematic prospecting demands a number of operations which, depending on the circumstances and the country in which the survey is conducted, require the services of 10 to 20 men:

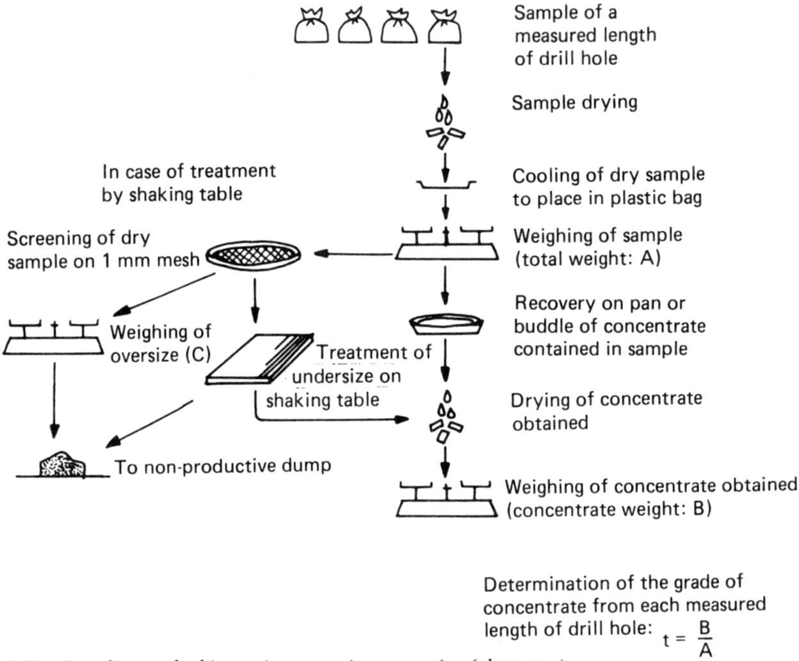

Fig. 3.20. *Sampling method in semi-systematic prospecting (phase two).*

1) educated laboratory manager for supervising storage, classifications, performing weighings and keeping sample records up-to-date etc.;
2) quarterer using Jones sample-dividers (or two medium-sized instuments and two small instruments) for dry quartering, or plastic containers for wet quartering (cakes);
3) 1 to 2 persons for supply/storage work;
4) 2 to 4 persons to dry the sandy material and concentrate;
5) 5 to 10 panners.

6.2.4.2 Preparation of composite concentrates

These samples must be prepared with the greatest care. Each composite sample includes a mixture of buddle or shaking table concentrates, obtained from the ungraded or raw ore from each measured length of mineralized drill hole perpendicular to the main axis of the rich zones of the deposit, with the ungraded material displaying a grade higher than or equal to the cut-off grade (eg 5%).

Lengths of lower grade than the cut-off grade are eliminated regardless of their position in the drill hole. Hence the number of composite samples will be identical to that of the drilling rows. It should be noted that the cut-off grade of 5% adopted in Australia is not necessarily the one to be used in other countries, where the geology is not comparable.

The composite may be obtained by mixing all the concentrates obtained on a profile if the drill holes have the same diameter throughout their length, which is not always the case. It is therefore important to hold a control sample of the concentrate of each measured concentrate of the same size to make up a composite sample.

To simplify the calculation of the quantity to be taken the Australians, large producers of ores from coastal deposits, have calculated a coefficient K called the density correction coefficient and have established a rating:

$$K = 1 + 0.007566t$$

where t is the grade expressed in percentage points of heavy minerals in the measured length of the drill hole (1 point = 1%).

6.2.4.3 Standard rating
For a 20.4% grade, for example, the coefficient K is:

$$1 + 0.151 + 0.003 = 1.154$$

For a 63% grade, for example, the coefficient K is:

$$1 + 0.378\ (50\%) + 0.018\ (13\%) = 1.396$$

The calculation basis of this coefficient rests on the fact that the density of mineralized sands depends on their heavy mineral content. To obtain a weight of each concentrate proportional to its extent the coefficients calculated as a function of t must be multiplied by an equivalent volume selected so as to obtain a composite sample weighing about 1kg or two composite samples weighing about 1kg each. A standard volume of 1 litre is often taken. **(Table 3.7.)**

Table 3.7. *Density correction coefficients.*

t = % HM	K = 1 +	t = % HM	K = 1 +	t = % HM	K = 1 +
0	0.000	11	0.083	31	0.234
0.1	0.001	12	0.090	32	0.242
0.2	0.001	13	0.098	33	0.250
0.3	0.002	14	0.105	34	0.257
0.4	0.003	15	0.113	35	0.264
0.5	0.004	16	0.121	36	0.272
0.6	0.004	17	0.128	37	0.280
0.7	0.005	18	0.136	38	0.287
0.8	0.006	19	0.143	39	0.295
0.9	0.007	20	0.151	40	0.302
1	0.008	21	0.158	41	0.310
2	0.015	22	0.166	42	0.317
3	0.022	23	0.174	43	0.325
4	0.030	24	0.181	44	0.332
5	0.037	25	0.189	45	0.340
6	0.045	26	0.196	46	0.348
7	0.053	27	0.204	47	0.355
8	0.060	28	0.211	48	0.363
9	0.068	29	0.219	49	0.370
10	0.075	30	0.227	50	0.378

Hence the weight of the concentrate to be sampled from a concentrate taken from 1m of mineralized sand at ti% is:

$$1000\ (1 + 0.007566\ ti) \times ti/100 = 10\ (1 + 0.007566\ ti) \times ti$$

6.2.4.4 Procedure
This example is more representative. Consider a row of pits with drill holes S1 and

S2 which have cut through a zone of grade above 5%. The two drill holes S1 (10m) and S2 (7m) have yielded the following heavy mineral grade, respectively (obtained by buddle or shaking table) from each measured length:

1) S1 over 5m (0 to 5m) : 9%, 13.50%, 20%, 35% and 45.08%;
2) S2 over 3m (0 to 3m) : 5%, 14.08%, 24.50%.

Taking the standard volume as 1 litre the weights of concentrate to be taken from each measured sample are the following:

Drill hole S1	0 to 1m	10 × 1.068 × 9	= 96g
	1 to 2m	10 × 1.102 × 13.5	= 148g
	2 to 3m	10 × 1.151 × 20	= 230g
	3 to 4m	10 × 1.263 × 35	= 442g
	4 to 5m	10 × 1.346 × 45.08	= 606g
Drill hole S2	0 to 1m	10 × 1.037 × 5	= 51g
	1 to 2m	10 × 1.106 × 14.08	= 154g
	2 to 3m	10 × 1.185 × 24.50	= 289g

Total weight of composite concentrate is 2013g.

Once the weights have been calculated, each of the buddle concentrates corresponding to a measured length of a drill hole is taken, and the weight of concentrate desired is sampled, by successive quarterings.

For example, assuming that the total weight of concentrate from a 2 to 3m length of drill hole S1 (weight B in the sampling scheme) is 560g. Of these 560g, 230g must be sampled by quartering. After thorough mixing these 560g are passed through the Jones sample-divider. The theoretical weights obtained at each passage are (in g):

560, 280, 140, 70, 35, 17.5, 8.7, 4.4, 2.2, 1.1

A sample of 320g is obtained by mixing 140g, 70g, 17.5g and 2.2g. However, it is ineffective to continue quartering the sample. After obtaining the 17.5g portion the 140g, 70g and 17.5g portions are added and the mixture weighed. Theoretically, this weight should be 227.5g. In actual fact, it may be slightly higher or slightly lower. It can be made up to 230g with concentrate taken from one of the sampler boxes. It may sometimes be necessary to move 1 to 2g instead of adding to reach the desired weight. It can be asserted that the addition or removal of 1 to 2g is extremely unlikely to alter the composition of the sample.

The content of the sampler box taken containing about 100g is preserved as a control of the measured mineralized length in the drill hole (in the present case, this could be the second box containing 140g).

Once the previously calculated quantity is taken from each measured length sample, everything is thoroughly mixed and, in the present case, the composite thus formed weighing 2013g is sent to the sampler, to prepare two samples weighing about 1006g, one of which is stored as a control and the second sent to the central laboratory (in the present case, the formation of two samples weighing exactly 1kg each, by removing the extra 13g, can be considered, without altering the nature of the sample).

The foregoing discussion is based on measured length samples from each drill hole. In the event that the drill hole lengths have different and random lengths, l_1, l_2, etc., the volumes of the samples taken will differ from each other by a quantity proportional to the lengths. The sample weights of each section to be sampled to make up the composite must be adjusted for the length of the corresponding section. If the sample sections are all the same length, but not 1m (eg 1.50m), there is no need to change any of the calculated concentrate weights in the foregoing example.

6.2.5 REQUEST FOR LABORATORY ANALYSIS

The mixed concentrates, prepared by the method described above, and despatched for analysis to the laboratories, are subjected to the following three main determinations:

1) determination of the mineralogical composition of the conentrate;
2) grade of each of the commercial minerals (ilmenite, rutile, zircon, monazite, cassiterite, etc.) contained in the concentrate, expressed as a percentage;
3) chemical analysis of useful minerals encountered to assess the value and to alculate the weights of simple elements contained in the deposit.

6.3 *Phase three: assessment of the prospected deposit*

The results of semi-systematic prospecting have led to the delimitation of a deposit of which it is now necessary to determine the extent and to calculate the weight of raw ore and heavy minerals.

6.3.1 DETERMINATION OF THE MEAN GRADE OF THE CONCENTRATE IN THE SECTION

The treatment of the sand samples serves to obtain the grade of the concentrate in each measured length of each drill hole in the section. These grades are assigned their corresponding place and entered on a profile at the correct scale based on the outline below. (These drill hole rows can be surveyed roughly in the field by the clinometer and their profile plotted on the graph; in most cases and for convenience the distance scales are different from the height scales.)

6.3.1.1 Observations

Different problems of choice arise in establishing mineralized profiles (**Fig. 3.21**) before proceeding with calculations. In the example selected, they have been resolved as follows:

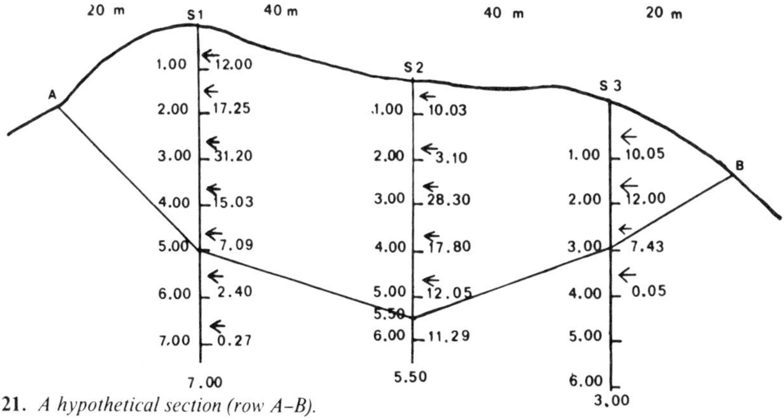

Fig. 3.21. *A hypothetical section (row A–B).*

1) On drill hole S1 the depth of the section was limited to 5m, because the underlying 2m contained only measured heavy mineral grades lower than the cut-off grade of 5%:
   — 5 to 6m: 2.40%;
   — 6 to 7m: 0.27%.
2) On drill hole S2 the measured section between 1 and 2m depth contained a grade of less than 5% (3.10%), but was surrounded by measured sections whose grades were greater than 5%:
   — 0 to 1m: 10.03%;
   — 2 to 3m: 28.30%, etc.

   It is obvious that when this section is mined the entire mineralized layer must be removed to a depth of 5.50m to remain commercial; a 1m thick deposit with a grade lower than the cut-off grade will not significantly reduce the mean grade of the section, which will be about 13.99%. Hence the 1 to 2m section containing 3.10% is included in the calculations.
3) On drill hole S3 the section between 5 and 6m contained a grade of 5.10% (greater than 5%) but it was separated from the upper mineralized levels by a 2m layer of less than 5% grade:
   — 3 to 4m: 0.05%;
   — 4 to 5m: 0.20%.

   It would not pay to extract these 2m of non-productive grade material to reach the layer between 5 and 6m. Hence the worked area will be limited to the depth of 3m. To calculate the mean heavy mineral grade (concentrate) in the section the data are entered in **Table 3.8**. The determination of other characteristics is shown below.

Table 3.8. *Mean heavy mineral grade data.*

Pit or drill hole No	Mineral seam length L	Heavy mineral content t	L×t	Mean grade Drill hole % $\frac{\Sigma L \times t}{\Sigma L}$	Mean grade Section % $\frac{\Sigma L \times t}{\Sigma L}$
S1	1	12.00	12.00		
	1	17.25	17.25		
	1	31.20	31.20		
	1	15.03	15.03		
	1	7.09	7.09		
	5		82.57	82.57:5=16.51	
S2	1	10.03	10.03		
	1	3.10	3.10		
	1	28.30	28.30		
	1	17.80	17.80		
	1	12.05	12.05		
	0.50	11.29	5.65		
	5.50		76.93	76.93:5.50=13.99	
S3	1	10.05	10.05		
	1	12.00	12.00		
	1	7.43	7.43		
	3		29.48	29.48:3=9.83	
Section 1+2+3	13.50		188.89		188.98:13.50=14

6.3.2 DETERMINATION OF DENSITY

Density depends on the heavy metal grade. Since the density of non-productive sand is 1.5, the density of mineralized sand is given by the empirical formula:

$$d = 1.5 + 0.0115\ t$$

where t is the mean grade, expressed in units, in the section considered. Hence for the section taken as an example:

$$d + 1.5 + (0.0115 \times 14) = 1.66$$

6.3.3 DETERMINATION OF THE MINERALIZED AREA IN EACH SECTION

The mineralized area in a section includes:

1) the sum of the areas of the trapezia whose bases are the mineralized lengths in two neighbouring drill holes and whose height is the distance between these drill holes;
2) outside the end drill holes (left of drill hole S1 and right of drill hole S3 in the example selected) the sum of the triangle whose base is the mineralized length in the drill hole (S1 or S3) and whose height is half of the interval between the drill holes in the section (40/2 = 2m in the example selected).

Thus in the example selected the area is:

$$50 \times 20/2 + (5 + 5.50)/2 \times 40 = (5.50 + 3)/2 \times 40 + 3 \times 20/2 =$$
$$50 + 210 + 170 + 30 = 460\text{m}^2$$

6.3.4 DETERMINATION OF THE VOLUME OF A DEPOSIT

The extent of each mineralized section whose area and mean grade have been calculated above is assumed to be equal to the half-distances between the section considered and each of the surrounding sections, regardless of whether they are

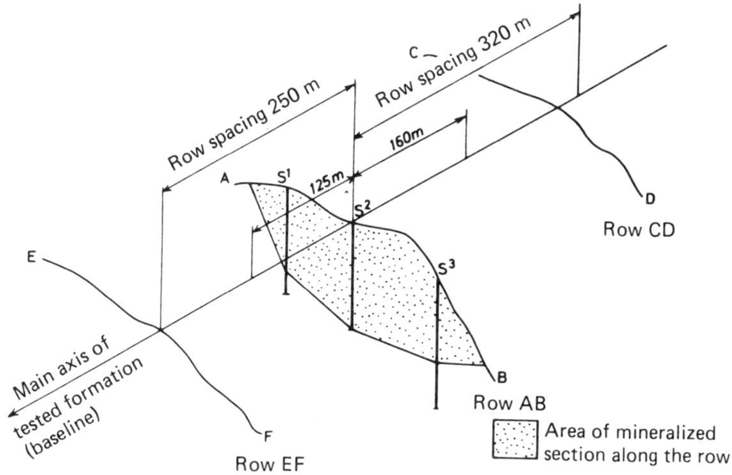

Fig. 3.22. *Determination of volume of deposits.*

mineralized or not. If the mineralized section is at the end of the axis, its extent is extrapolated by a length equal to half the distance between it and the previous section.

In the example above the volume of ore determined from the mineralized section AB, whose mineralized area is 460m^2, is **(Fig. 3.22)**:

$$460 \times (125 + 160) = 460 \times 285 + 131{,}100\text{m}^3$$

6.3.5 WEIGHT DETERMINATION

Three types of weight estimates can be calculated:

1) The weight of ungraded ore is obtained by multiplying the volume by the density. In the sample selected the weight of ungraded ore is:

   $131{,}100 \times 1.66 + 217{,}626$ rounded off to 217,600t

2) The weight of concentrate is obtained by multiplying the weight of ungraded material by the mean heavy metal grade (concentrate) of the section:

   $217{,}400 \times 14/100 = 30{,}436$t

3) The weight of each of the commercial minerals encountered in the deposit (ilmenite, rutile, zircon, monazite, cassiterite etc.). The grade of each of these commercial minerals is provided by the laboratory which has analysed the composite samples from each drill row. In the example selected the hypothetical section row AB has the following values supplied by the laboratory:

   Total percentage of ilmenite 80%
   Total percentage of zircon 9%
   Total percentage of monazite 3.25%

All these data are entered on a form which summarizes the results obtained on the drill row, which is presented as shown below. Care must be taken to enter the number of each row at the top, in the right hand corner of each sheet (R A–B in our example), in order to facilitate the search for a row by flicking through the collected sheets. The number of sheets of this type is equivalent to the number of drill rows (sections) along the deposit. The sum of the data obtained by the calculation of all these combined sheets gives the overall information on the prospected deposit in the second phase.

1) Mean grade

Section form

Pit or drill hole no	Mineralized length L	Heavy mineral grade t	L × t	Mean grade	
				Drill hole (%) $\frac{\Sigma L \times t}{\Sigma L}$	Section (%) $\frac{\Sigma L \times t}{\Sigma L}$
S1					

2) Density (specific gravity)
   $1.5 + (0.0155 \times 14) = 1.66$

3) Mineralized area of section

$$\frac{5\times 20}{2} + \frac{5+5.50}{2}\times 40 + \frac{5.50+3}{2}\times 40 + \frac{3\times 20}{2} =$$

$$50 + 210 + 170 + 30 = 460\text{m}^2$$

4) Volume of ore surrounding the section
— row spacing on one side of section: 250m
on the other side of section: 320m
— Volume = 460 × (125 + 160) = 460 × 285 = 131,100m²

5) Weights

Raw ore	131,100 × 1.66	217,600t
Concentrate	217,600 × 14/100	30,464t
Commercial minerals		
— Ilmenite	30,464 × 80/100	24,371t
— Zircon	30,464 × 9/100	2,742t
— Monazite	30,464 × 3.25/ 100	990t

6.4 *Phase four: systematic prospecting*

This is the final phase of prospecting during which all the data necessary for the feasibility study and for mining of the deposit must be gathered together including:

1) precise delimitation of the deposit and of its richest parts;
2) level of the water-table and its height above sea level;
3) section of substratum topography, and its state of weathering;
4) particle size distribution of the ore; note that this analysis is often undertaken in phase two as soon as the drill holes indicate the existence of an interesting deposit.

As in phase two the same methods for obtaining and for analysing the results on the basis of drill holes are employed in systematic prospecting.

6.4.1 PARTICLE SIZE ANALYSIS

This consists of taking, from the overall deposit tested, a series of samples in the richest parts intersected by the drill holes. These samplings are distributed among the drill holes of a section transverse to the main elongation of the sandy formation to be tested, or in drill holes carefully selected from the overall deposit tested. This analysis, which deals with ungraded ore and concentrates separated by heavy liquids (generally bromoform, specific gravity 3.3), is designed:

1) to determine the particle size classes in which the commercial minerals are quantitatively situated, and, from all the samples examined, to determine whether classes with stronger mineralization do or do not exist;
2) to determine the extent of particle size differences between the non-productive sand and the heavy minerals;

3) to determine whether the non-commercial heavy minerals (eg garnet, staurolite) are situated preferentially in one particle size range, and if so, which one?

All these data are necessary to programme the gravity concentration tests which will serve to design the treatment plant, but may also provide useful information on the deposit such as:

1) possible variations in particle size distribution between different dune ranges of a single, deposit;
2) the eolian character of the sandy formation, the richest parts, generally located in the upper levels of the formation being the finest, and the coarse sands which are little mineralized, if at all.

This analysis is conducted on each sample using screens with the following mesh sizes:
1, 0.400, 0.250, 0.160 and 0.080mm

Noted in each particle size class are the weight percentages of the commercial minerals collected after separation from bromoform (ilmenite, rutile, zircon, monazite, cassiterite, etc.) in each sample. As a rule, this sort of analysis can only be performed in a specialized laboratory. The results of this analysis are summarized in tables of which one example is given below (**Table 3.9**).

Table 3.9. *Example of synoptic table of granulometric analysis.*

			Heavy minerals after bromoform separation		Ilmenite		Zircon		Monazite		Other remaining unquantifiable heavy minerals	
			Weight (g)	%	Weight (g)	%	Weight (g)	%	Weight (g)	%	Weight (g)	%
A - Deposit X (West) **Row R14** **Borchole S.222 (0 to 3m)** Total: 1177.5 g												
Ø 1 m/m	58 g	4.93 %	0	-	0	-	0	-	0	-		
0.400	625 g	53.15 %	0.10	0.64	tr	-	tr	-	0	-		
0.250	386 g	32.82 %	1.30	8.31	0.80	6.32	0.20	16.67	tr	-		
0.160	69 g	5.87 %	8.05	51.44	6.80	53.75	0.50	41.67	tr	-		
0.080	26 g	2.21 %	5.85	37.38	4.80	37.94	0.50	41.67	tr	-		
< 0.080	12 g	1.02 %	0.35	2.24	0.25	1.98	tr		tr	-		
	1176 g	100 %	15.65	100	12.65	99.99	1.20	100	1.20	-	4.20	0.30
B - Deposit X (East) **Row R14** **Borchole S.222 (3 to 8m)** Total: 1230 g												
Ø 1 m/m	294 g	23.90 %	0	-	0	-	0	-	0	-		
0.400	666 g	54.15 %	0.05	0.68	tr	-	tr	-	0	-		
0.250	151 g	12.28 %	0.35	4.79	0.15	3.30	0.05	9.09	0	-		
0.160	61 g	4.96 %	1.70	23.29	1.00	21.98	0.10	18.18	tr	-		
0.080	46 g	3.74 %	4.25	58.22	2.70	59.34	0.30	54.55	0.40	80.00		
< 0.080	12 g	0.98 %	0.95	13.01	0.70	15.38	0.10	16.18	0.10	20.00		
	1230 g	100 %	7.30	99.99	4.55	100	0.55	100	0.50	100	1.70	0.12

Appendix 1

Characteristics of the drilling equipment used in prospecting littoral formations

1 Introduction

The drilling equipment used for prospecting littoral formations over the last 30 years has comprised simple manual or motorized drills. These are:

1) manual drills, which drill non-cased holes and consequently are used to investigate the shallow water-table;
2) 'Banka' type drills which can be manual or motorized, and are able to penetrate the the water-table and reach the bedrock below.

Both drills have advantages and disadvantages. Australian mining companies have mastered the art of evaluating littoral mineralization using a drill that combines the advantages of the two types of apparatus and eliminates certain disadvantages. This equipment combines lightness and robustness with a speed of operation which increases output of the best drill currently used five-fold, whilst permitting the water-table of the prospected formations to be traversed.

These different models of drill are described below and their advantages and disadvantages are presented.

2 Hand auger drill ('Helix' type)

This type of drill is simple but does not give cased drill holes: the well sunk by this equipment will stop as soon as the water-table is reached and will not be able to go beyond that point. Therefore, it is a tool which is somewhat limited but it can be used for reconnaissance prospecting.
This equipment includes:

1) one helicoidal tool, which is used for collecting the sand from the well by rotation (this tool is available in 60 to 70mm, 80mm, 100mm and 150mm cross-sections and the first two sizes are the most commonly used);
2) one drill string of five rods having a minimum length of 2m;
3) one rod 1m in length;
4) one tap wrench;
5) two spanner rods;
6) one stopping fork for the drill string;
7) two salvaging bowls for the collected sand;
8) one bucket with a 15 to 20 litre capacity;
9) one wooden basin (made locally);
10) two wooden blocks to hold up the stopping fork of the drill string;
11) X sample bags of 30 x 40cm made of plastic (on average between one or two bags of this size are needed for each metre being drilled);
12) possibly two small canvas sheets for on the spot coning and quartering.

In desert countries, a canvas shelter must be provided as protection from the wind (at ground level) and sand which is likely to contaminate the samples during processing operations. The workforce necessary for the handling of this equipment is a minimum of three people per drill (four being the maximum). The rate of advance per team is from 20 to 30m per 8h shift, according to the terrain. As the drilling is carried out without well casing the risks of contamination from the sand off the well is very high; therefore, the surface of the tool must be cleared by hand each time the product is recovered and only the core of the sample brought up on the tool must be retained.

ADVANTAGES

1) simplicity of the apparatus;
2) robustness (made of steel);
3) easy manipulation;
4) small workforce (a maximum of four people);
5) a very quick period of instruction for operators in the use of the apparatus;
6) capability of reaching locations with difficult access;
7) a depth exceeding 20m in homogeneous sandy formations, slightly humid (naturally or artificially);
8) ease of measuring the depths that have been reached.

DISADVANTAGES

1) Although it is composed of detachable components, the equipment is relatively heavy. Recovery of the drill string (by hand) becomes more and more difficult as the well deepens.
2) There are high risks of contamination from sand off the well walls since the latter is not cased.
3) During deep drilling the weight of the string is likely to break the blades off the salvaging tool.

As the observation wells are not cased drilling will be stopped as soon as the water-table has been reached. Hence this equipment does not permit the levels underneath the water-table to be examined.

3 Light-weight drill ('Banka' drill)

While reconnaissance prospecting of littoral formations does not in principle require drilling below the water-table, in semi-systematic prospecting, it is essential to continue drilling as far as the bedrock by traversing the water-table or at least as far as the substratum of the sandy formation. The hand auger drill used in reconnaissance prospecting must therefore be replaced with a piece of drilling equipment which enables drill holes to be cased.

The equipment sometimes used for this type of prospecting is the 4 inch 'Banka' drill. This type of drill has been developed to sample alluvial formations of heterogeneous grain size. It can be used in more homogeneous littoral formations but it is less useful because of its weight and low output.

4 Light-weight Australian drills

INTRODUCTION

The problem of prospecting littoral formations has been studied by different Australian mining companies who have perfected light-weight drills varying from the simplest to complex types. Three light-weight hand drills are described and these can be activated generally by backup motors independent of the drill string.

In light alloy (steel or aluminium) these drills generally comprise a core drill (sand pump) with equipped in its lower section a bit with a clock valve and in its upper section with a bevelled opening, allowing sand to be recovered by turning the apparatus upside down. This drill, mounted on a drill string, is driven into the ground by percussion or rotation under constant water injection so as to avoid the sand rising.

These drills require a small workforce: a team of a maximum of 4 or 5 men drilling holes of 6 to 7m in depth can carry out about 40m of drilling per 8h shift according to the terrain.

'HAND AUGER DRILL KIT' (PORTABLE NON-CASED)

This small drill consists of the following components allowing a well of 6m to be sunk:

1) one auger of ø 5cm;
2) one corkscrew of ø 5cm;
3) seven drill rods in aluminium (1m in length);
4) one tap wrench;
5) two spanners for assembling-dismantling;
6) one case with a handle for transporting the apparatus.

Altogether this equipment weighs 12kg. It has the advantage of being able to be stored in a bag and transported like a briefcase.
It enables:

1) The collection of samples from any formation (ground, sand or clay) is very fast. Its cost is slightly higher.
2) Although the drill can be operated manually, a separate motor can be attached to it.

HAND DRILL (THE ONE USED MOST FREQUENTLY FOR LITTORAL FORMATIONS)

In order to be able to continue the drilling below the water-table the bore hole must be cased. Additional components whose features are given below must be attached to the equipment described above. A boring, technique developed by the Australians called 'below the water-table' allows drilling to be continued under water after having traversed the water-table.

Characteristics of the non-cased type
In order to be able to continue the drilling below the water-table the bore hole must be cased. Additional components whose features are given below must be attached to the equipment described above. A boring technique developed by the Australians called 'below the water-table' allows drilling to be continued under water after having traversed the water-table.

Table 1. *Characteristics of the non-cased drill.*

Components	ø in cm*	Length in m**	Weight in kg	Observations
Sand auger	3.75	1.50	2.00	
	5.00	1.50	2.00	
	6.25	1.50	2.50	
	7.50	1.50	2.50	
	10.00	1.50	3.00	
Screw auger	5.00		2.00	
Drill rod	3.75	1.50	1.00	
alluminium	2.80	1.50	0.75	
steel	2.80	1.50	1.50	The stainless
Drill tube			1.50	steel drill
(aluminium)			0.25	rods are more
Star bit			2.50	economical
				and more
Tap wrench			0.50	resistant
Spanners (pair)			0.50	
Total: in aluminium		1.50	22.0	
in stainless steel		1.50	22.75	

* Hand drills have different ø whose size varies according to the depth of the bore hole.
** The lengths given here have been converted from Imperial units to metric units.

Table 2. *Characteristics of the 'below the water-table' cased type.*

Components	ø in cm*	Length in m**	Weight in kg	Observations
No 1 Tube	5.00	1.50	3.50	The drill string
	6.25	1.50	5.00	and the recovery
No 2 Sliding shuttle	5.00		0.50	equipment (sand
for the tube	6.25		0.50	pump) exist in
No 3 Sand pump	3.75		1.00	different lengths
(on cord)	5.00		1.50	- here we have
No 4 Extendable	3.75		1.50	chosen the one
sand pump	5.00		2.00	closest to the
No 5 Nozzle (part)	3.75		2.00	metric system
	5.00		2.50	where 5
				inch equals 12.7 cm
No 6 Bevelled	3.75	1.50	1.50	
sand pump	5.00	1.50	2.00	
No 7 Rotating belt			1.00	
and cradle			–	
No 8 Platform	0.60 × 0.75		6.00	
Total		1.50	30.50	

Modes of use

When the first drill tubes have been inserted in the well after core removal has been made by the auger drilling is carried out by a platform-cramp iron held tightly on the end of the drill tubes in order to turn the drill string. It also uses the platform notably when it is being worked with a rope. The system of drilling using a rope to replace the drill string is not advisable due to the possibility of the rope getting twisted during use or broken, etc.

Output

The drilling progress is spectacular.

1) This type of auger goes down about 0.30m with each rotation of five turns.
2) One man (or more efficiently two) can sink 9 to 10m into homogeneous sandy formation in 1.5h.
3) A team of three experienced men can drill 18 to 20m in 1h under the same conditions; with an inexperienced drill crew the same depth can easily be reached in 2h maximum.
4) A team of three men working for 8h can easily drill to a depth of 48 to 50m of non-consolidated sand, including transfer on a row of wells whose inter-well distance is anything up to 40m.

Finally, it must be noted that this type of auger can not be used in a heterogeneous formation made up of coarse material whose constituent elements have a diameter greater than 3.7cm (eg gravel).

This hand drill is an ideal tool for prospecting sandy littoral marine formations and notably for those which have been reworked by the wind. When the wells sunk with this auger are not cased the drilling being carried out by means of this equipment will not, as a rule, cross the water-table. However, this highly developed form of the helical tool for sand recovery has enabled the continuation of the drilling without casing to more than 2m below the water-table.

The sand pumps can be used fastened to a rope (No 3 or No 6) (which is not recommmended) or screwed on to a drill string (No 4 and No 6). There are three types of sand pump:

1) Rope sand pump (No 3): this is the simplest, it is supplied with a grab handle at its upper end to which the rope is attached.

2) Extendable sand pump (No 4): this model is heavier.
3) Bevelled sand pump: this combines the two pumps described above and offers several possibilities.

It is a piece of equipment which can be used either on an extension (No 4) or with a rope (No 3). The sliding shuttle is joined to the pump at its lower end and the emptying is done through a bevelled hole on the side. It is available in a 3.75 or 5cm slashed O and can reach 1.50m (metric equivalents).

A tube of 6m can be withdrawn in one go by hand after the drilling has finished (otherwise a jack is used at the beginning) to free the equipment. During drillings extending as far as the bedrock, after crossing 21 to 15cm of a very compact substratum, you may have to resort to using a screw jack to extract the drill string!

The screw jack is not found in the Australian equipment inventory because the drill is only intended for boring into sandy formations.

Three men can drill through the water-table 6m deep in 1h and sink a hole as much as 15m below the water-table.

The only difficulties encountered will be due to sand rising; it is essential to fill the bore hole with water so as to pressurize the tubing. A funnel on the end of the last tube is provided for this purpose.

Disadvantages

During use the components, such as the tubes and augers, are subject to wear and tear. The rate of deterioration, just to give an approximate idea, will be the following (in a sandy terrain of homogeneous grain size):

1) one tube in seven is worn out after a fortnight in use;
2) one auger is worn out after 2km of drilling.

'Drillmite' motor

This motor runs on petrol and the controls coupled to the drilling instruments are hydraulically driven. The rotation is controlled by hand. The total weight of the motor and hand drill ('hand auger drill kit') is 50kg. The entire equipment can easily be worked by two men and loaded on to a cross-country vehicle.

'SANDRILL'

Features

This drill presents numerous advantages. The main advantage is its simplicity. It is not necessary to engage specialized personnel for its use. It is composed of a helical screw auger with renewable cutting edges which by continual rotation brings the sample to the surface. The entire equipment is powered by a 10HP motor which powers a hydraulic pump coupled to the motor which includes a water tank, filter, pressure gauge, flexible pipes and quick release joints. It weighs 400kg and it can be transported by a cross-country vehicle, eg a Landrover, with the drilling team. The motor lifts or lowers the drill string by a system driven by a rubber wheel. The tubes are made of aluminium, with a second tube inside mounted on ball bearings measuring 1.5 or 3m and weighing 3kg.

This drill is mainly used for drilling non-consolidated sand; it should not be used for sedimentary formations containing gravel and detrital rock constituents.

It is in this category that we find the most economical use of all drills of this type, the lightest and that which presents the best output when set up. For example:

1) to drill 60m the 'Sandrill' takes 2h;
2) to drill 60m the 'Banka' drill takes one week (one 8h shift per day).

Chapter 4

Geochemical prospecting

1 Basic ideas

1.1 Definition of geochemical prospecting

Geochemical prospecting entails the systematic measurement of one or more trace elements in rocks, soils, stream sediments, vegetation, water or gases. The aim is to reveal geochemical anomalies, ie abnormal concentrations of certain elements which clearly contrast with their environment or 'geochemical background'. The formation of anomalies results from the mobility and dispersion of elements concentrated in mineralized zones.

An anomaly can be defined as having a higher or lower (negative anomaly) content than the geochemical background. Its origin is not exclusively metallogenic but can result from foreign contamination such as copper pollution by vineyard sulphatization or contamination by the waste products of old or recent mining work. In addition, there are anomalies termed 'formational' caused by an element preferentially linked to a particular geochemical formation (Cu in basic rocks, Pb-Zn in dolomites). However these anomalies have a particular mineralogical form (Cu bearing silicates) or are connected with a particular type of deposit (finely dispersed) so they are not of commercial interest. One of the geochemist's main objectives is to distinguish between a 'real anomaly and a 'formational' or 'pollution' anomaly.

Two kinds of dispersion can be considered: (1) a primary dispersion connected with the emplacement phenomena governing the mineral concentration, eg in hydrothermal alteration. The study of primary deposits thus formed is useful in prospecting for deposits and is done by sampling (drill hole cores, cuttings). The major elements called 'tracers' are either chosen from elements constituting the host rock or from metallic mineralizing elements. It is then possible to trace deposits and to determine dispersion gradients and directions in the vicinity of a deposit; (2) secondary dispersion linked to weathering and geomorphological phenomena **(Fig. 4.1).**

Such mechanical and chemical dispersion, from the destabilized metal stock into the oxidation zone, causes deposits and secondary dispersions which cover a greater area than the intersection by the erosion surface of a mineral concentration, hidden by various cover materials. Geochemical prospecting determines these secondary deposits by the chemical analysis of soil samples, stream sediments or altered rock.

Different elements found in a deposit may have totally different mobilities and hence may not give spatially similar anomalies. For example: lead and zinc are very

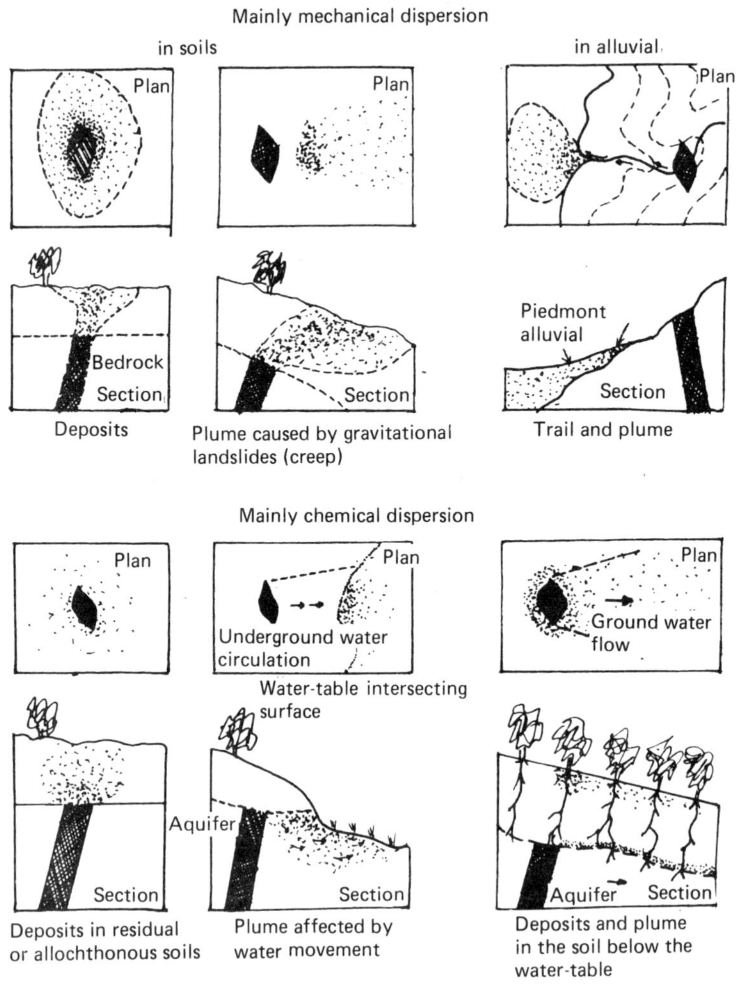

Fig. 4.1. *Secondary dispersion and deposits displaced from their source.*

frequently associated in ore deposits, yet in a silico-aluminous environment, under superficial weathering conditions, zinc is very much more mobile than lead. The zinc is much more likely to be leached whilst the lead would give an anomaly in the mineralization.

The geochemistry of a secondary anomaly is not a direct reflection of the geochemistry of an underlying deposit, but the result of the combination of the following phenomena: primary geochemistry; oxidation and mobility conditions and the effects of the secondary environment. Care must be taken not to conclude that an ore deposit exists just on the basis of an 'anomalous paragenesis' (secondary concentration).

The approach discussed is soil, stream sediment or rock sampling, since these are by far the most common approaches. However, other approaches exist and should be considered in particular situations: the hydrogeological approach using water analyses (this is restricted to analytical detection limits – elements present as a few ppb – it is currently used for fluorite and uranium); the phytogeochemical approach (the analysis of ashes after burning certain vegetable matter) and the analysis of panning concentrates (a method especially suited for prospecting certain metals).

The remainder of this chapter concentrates on the practical ways for the use of different methods and where they should be applied.

1.2 Progressive phases of development and aims

Geochemical methods may be used in all the various phases of prospecting: involving the search for secondary anomalies; general survey and strategic prospecting. Finally, geochemistry can still be useful for the exploration and the development of mineral occurrences (study of primary deposits).

General regional surveying consists of covering a relatively unknown region by widely spaced geological traverses to collect the following information: geographical data (drainage course, relief, road and track network); geological data (type of formations found: sedimentary or basement, etc.); pedological data (alteration type, nature of soils, the presence of stream sediments, etc.). Sampling along the traverse and exploratory cross-sections is neither systematic nor completely random. As far as possible these cross-sections should be made perpendicular to the general lie of the formation, and samples should be collected consistently throughout the cross-sections so as to reveal variations in the geochemical background. The purpose of this regional approach is to reveal the principal geochemical elements. It leads to the selection of zones to be prospected by semi-detailed geochemistry and to the methods to be employed.

Strategic prospecting is a systematic approach which should provide continuous information throughout the prospect. Its aim is to reveal the anomalous zones probably connected with mineralization in any given region.

Tactical prospecting is also a systematic method. By exploring zones revealed at the strategic prospecting stage it examines the origin of these anomalies, their superficial extent and their mineral grades. At each of these three stages, it should be carefully noted that using surface information only gives an approximate picture of the intersection of the deposit with the alteration surface: secondary anomalies. In certain cases, detailed geochemistry could be completed by some very local spot studies of surface exploration, eg geochemistry on auger samples and trench analyses.

2 Geochemical prospecting planning programme

According to the programme's objectives and the prevailing climatic conditions, a proposed geochemical prospecting campaign should be planned with the aid of a consultant geochemist. Three main points should be examined: type of sampling, density and elements to be analysed.

2.1 Type of sampling

As stated above, samples will be taken from stream sediments, soils or from rocks. For each of these three types a methodological study would normally be made in order to define the best grain size fraction for analysis, or to see whether there exists any metal bearing phase (hydroxides, oxides) which could be isolated and whose analysis would reveal mineral grades and higher geochemical contrasts (anomalous values/background values). This study could be done at the regional survey stage where all types of samples are tested. In the case of semi-regional prospecting stream sediments are most usually sampled, a technique which nowadays seems to be the most suitable. However, for various reasons some of the samples may have to be soils. It will therefore be necessary to increase the sampling density and collect the soils from low areas rather than in crest zones. For certain metals, particularly those of which the minerals are in the form of oxides (W, Sn), the analysis of panning concentrates combined with an optical examination provide greater precision and accuracy.

Tactical prospecting is based on soil sampling normally at the base of level A **(Fig. 4.2)**. The presence of a particular type of weathering, lateritic duricrust or allochthonous cover may require special sampling techniques, eg sampling from a coarse soil fraction in deserts with possible aeolian cover or auger sampling in level C in the case of clearly allochthonous cover.

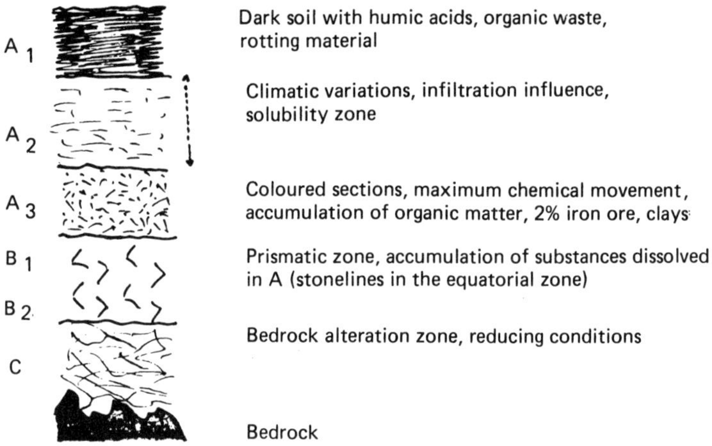

Fig. 4.2. *Ground in cross-section.*

2.2 Sampling density

Sampling density varies according to the scale of prospecting and the aims of the survey. In strategic prospecting a map scale between 1:200,000 and 1:50,000 is generally used. For stream sediments sampling is carried out on a large grid with a density of one to several samples per km^2. The idea of sample spacing along the network is not so important as the idea of average density, since this will provide representative information. In the event of combined soil and stream sediment sampling, the density should be increased. In any case, strategic prospecting should not fall below one sample per km^2. Such a grid should be aimed at accurately defining anomalies related to mineral concentrations. This density also provides a satisfactory picture of the local geochemical background. All anomalous points, even if isolated, should be taken into consideration.

At the detailed stage where sampling is carried out on a regular equally spaced network one no longer refers to density per km^2 but to a grid. The latter will vary according to the mineral sought. A 200 × 200m grid is normally sufficient for a first location of Pb, Zn or Cu anomalies but a 50 x 50m grid will be required for Sb or W anomalies. A tightening later is always preferable to initial 'over gridding'.

2.3 *Analysis*

This is an essential stage; the choice of one technique in preference to another could totally change the efficiency and the significance of the proposed programme. The technique employed should at the same time be reproducible, sensitive and economical. Two main approaches are possible: simple methods which can be employed in the field, for example as in a laboratory van; or more sophisticated techniques, which are carried out at a central laboratory. Both have their advantages and disadvantages. There are a number of laboratories which commercially carry out geological prospecting type analyses, where analytical work can be subcontracted. However, one essential precaution would be the introduction of standard samples with known concentrations in order to verify both the reproducibility and the quality of the analyses provided by the laboratory.

At the BRGM the application of multi-element analysis techniques has modified the selection criteria; it is becoming more and more common at the strategic prospecting stage. At the tactical prospecting stage the problem is usually much more clearly defined and only a small group of elements will be analysed. However, some elements other than those which are directly sought are very useful for judging whether or not the anomalies are of a metallogenic type. For example, the Ni content is systematically analysed in tactical prospecting for Cu; in the same way, As, Ag, Mo, Cd are useful guides for the prospecting of Pb–Zn.

When mineral occurrences are being explored – and especially in the case of the examination of primary deposits by rock geochemistry – multi-element analyses are indispensable. The variation in the concentration of the major elements is of prime interest, as are the composite metallic rings which, in some cases, will show the orientation and the extent of erosion in the deposit.

In studying gossans it will be on the basis of elements such as Sn, Bi, Sb, Ag, Mo, etc. that the value of the primary mineralization can be assessed. In some instances even trace analytical techniques should be used for Au, Hg, etc.

2.3.1 THE MOST COMMONLY USED ANALYTICAL METHODS

At the BRGM two multi-element analytical techniques are currently used:

1) Plasma emission spectrometry: 12 priority elements (Cu, Pb, Zn, Ag, W, Sb, Ba, Ni, Mn, Fe, Cr, Sn) and 10 useful elements either as accompanying elements or for geological mapping: V, P, As, Mo, B, Be, Cd, Co, Ni, Y.

2) Direct reading optical emission spectrometry: quantometer, which simultaneously measures 7 major elements and 26 trace elements (Si_2, Al_2O, Fe_2O_3, MgO, CaO, Na_2O, K_2)–Mn, P, Ti, Zr, B, Ba, Sr, La, Y, Nb, Pb, Zn, Cu, Ag, Cd, As, Sb, Bi, Li, Sn, W, Mo, Cr, Co, V, Ni.

The atomic absorption technique is the most used in standard single element analysis.

2.3.2 IMPORTANT RECOMMENDATION

The various analytical techniques have distinct sensitivities according to the metals analysed and it is absolutely impossible to compare results provided by different techniques. The same analytical technique should therefore be employed throughout any semi-detailed campaign. Finally, so as to ensure analytical reproducibility and accuracy, duplicate or reference samples should be inserted, at the rate of approximately one sample per 100 analysed. The duplicate check samples are used as internal standards. Some numbers should be left without samples so that the reference samples can be inserted.

3 Setting up the various stages of prospecting

Once the objectives and the overall concept of the expedition have been defined with the help of a consultant geochemist, it is up to the prospector to carry out the field work. Whatever the prospecting stage, the work comprises route and location of sampling, collecting and numbering of samples and preparation and despatch of samples.

3.1 The route and location

At the general regional survey stage, so as to cover the largest possible surface in a minimum amount of time, it is important to make the most of natural or man-made means of communication, where possible. For example, small boats could be used on navigable waterways, with compass traverses on dry land between the various drainage courses. In most cases, existing topographical maps, at a scale of 1:100,000 or 1:200,000, will suffice for the location and positioning of sampling points (crossings of the paths and drainage course and distances along the compass traverses between them). In the strategic phase the aim is to locate anomalies on which more detailed prospecting can be concentrated later; the sampling density and its regularity are of prime importance. The positioning of the sample site should also allow for a safe rapid return to anomaly points. Stream sediment sampling is the most usual method.

Thus the most logical route is marked out along the secondary water course. In order to maintain a constant sampling density, it is advisable to preselect sampling points on a topographical map or on available aerial photographs, superimposed on a kilometre grid. Density should always take precedence over spacing along the length of the network. Moreover, these two notions are not always independent. For example, experience has shown that in the equatorial forest zone (eg Gabon, Zaire) one sampling per kilometre of a water course approximately represents a density of one sampling per km^2; in France, a sample taken every 250m corresponds to a density of 8 samples per km^2 **(Fig. 4.3)**.

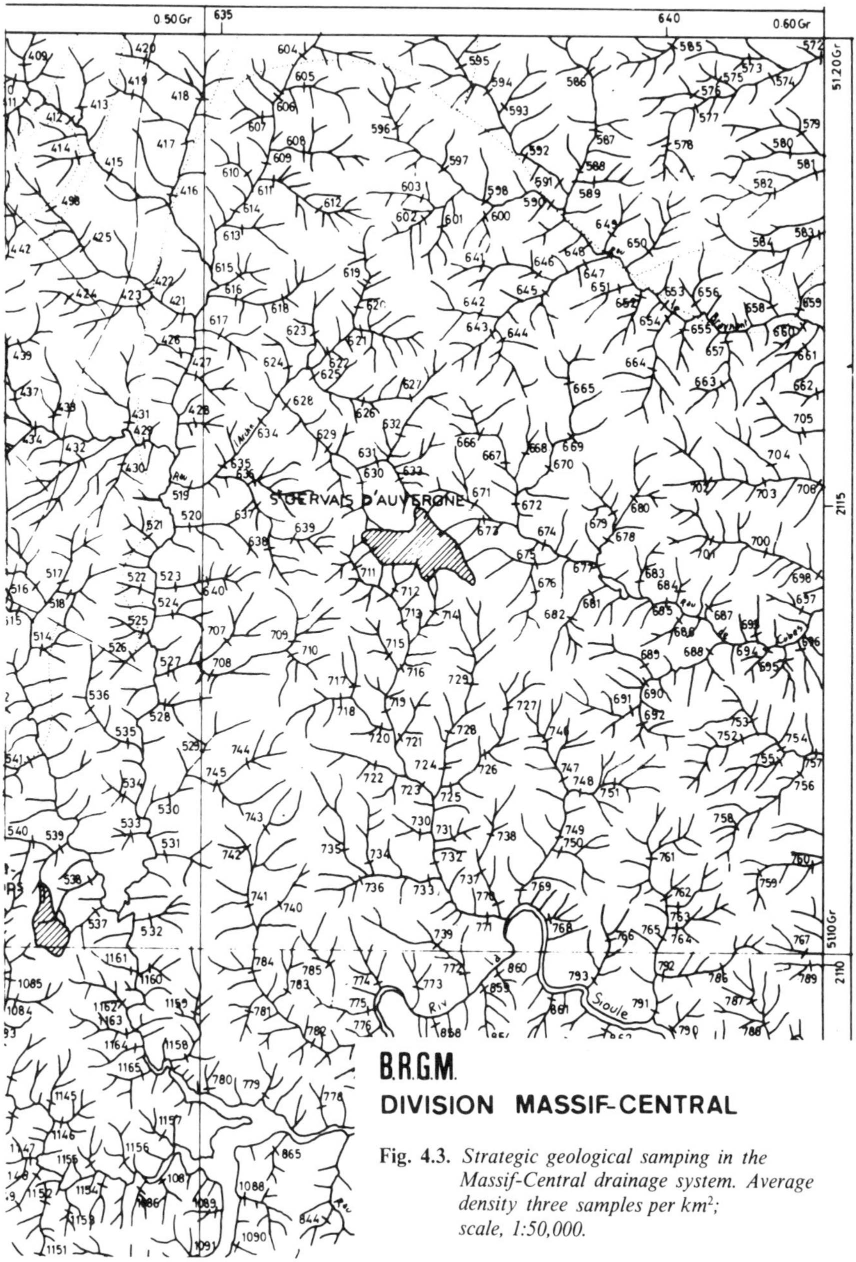

Fig. 4.3. *Strategic geological samping in the Massif-Central drainage system. Average density three samples per km^2; scale, 1:50,000.*

Exact positioning is made with the help of traverse surveying (compass and 'topofil') and sampling points plotted on the topographical map. Most important is to be able to find the sampling point again without difficulty. Should the location of sampling points in the drainage course prove too difficult, then sampling rows can be set across the waterways. However, in many cases too few waterways are intersected and it is necessary to add soil sampling on a regular grid to the stream sediment sampling carried out in each of the intersected waterways. For example,

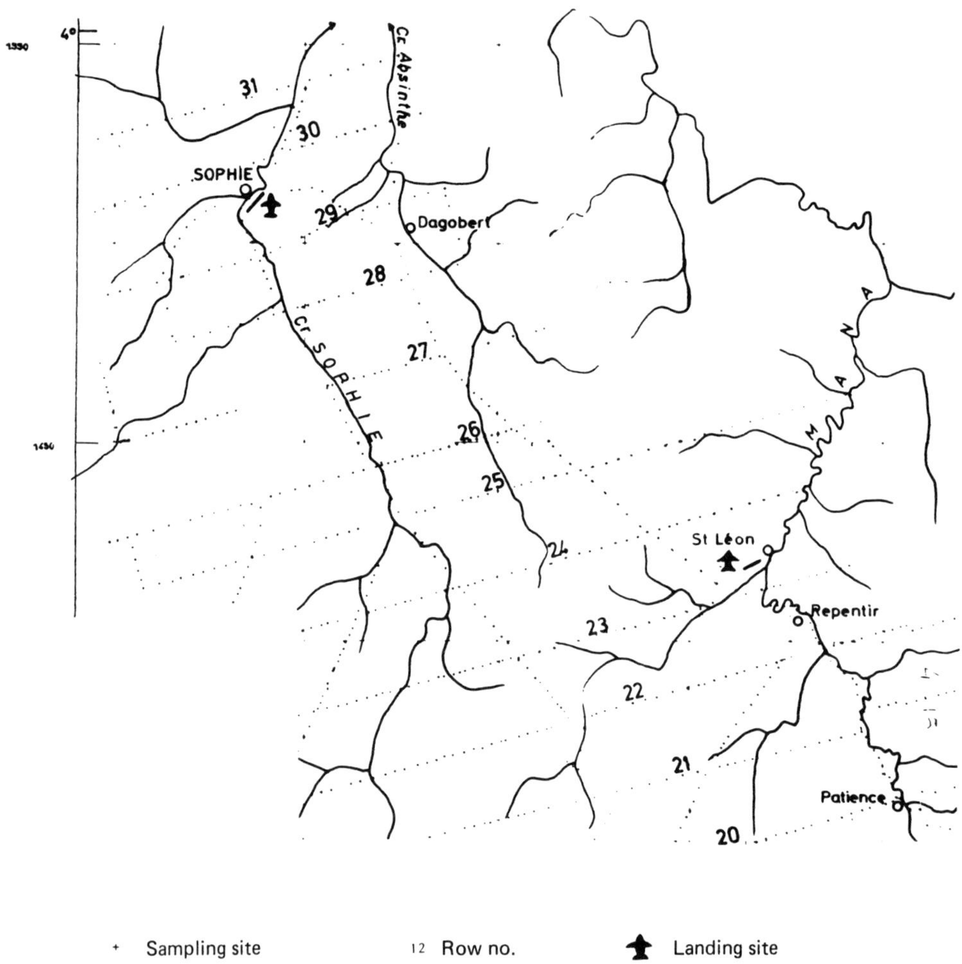

Fig. 4.4. *Map of strategic geological sampling in Guyana. Sampling is carried out in rows 2km apart with one soil sample taken every 500m plus one stream sediment sample taken from each water course crossed. Density, two samples per km²; scale 1:10,000.*

at a 1:100,000 scale, rows spaced 2km apart with sampling made every 500m (grid used in Guyana) produce one soil sample and 0.5 stream sediment sample every km^2. This combined soil-stream sediment density is just as representative as a stream sediment density per km^2. It is always preferable to take as many stream sediment samples as possible and modulate soil sampling according to the lie of the land, and that samples are from the valley slopes or floor to crest **(Fig. 4.4)**.

With the aim of strategic prospecting being the location and assured relocation of anomalies it is imperative that each sampling point is marked in the field. In the forest a plastic strip numbered with a bright indelible marker other than green

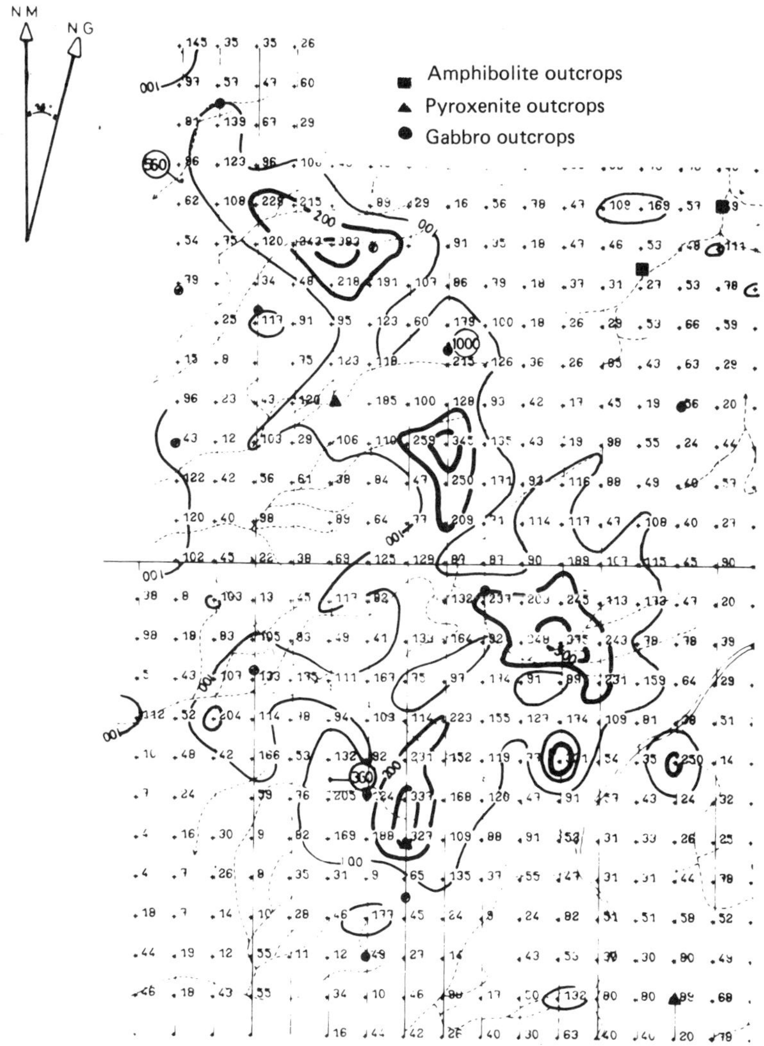

Fig. 4.5. *Tactical prospecting grid for Cu. Square grid, 200m.*

seems to be the most efficient way and can be found again several years later. In a desert area a paint mark on the nearest outcrop is often a good solution. The above envisaged route is a walking route, often the only possibility in forest areas. In other conditions various other routes are possible: mopeds or motorbikes can be used in savannah where there are many tracks, motorbikes or even a car can be used in desert areas and a helicopter can be indispensable sometimes in mountainous regions. Whilst speeding up sampling, each of these forms of transport presents a particular location problem, which will have to be solved in each case.

Aerial photographs can determine the best means of locomotion. Satellite images can often be enlarged which may prove useful at this stage. Logistic problems such as the organization and scheduling of flights, kerosene supplies, etc. can sometimes determine the choice of transport and should therefore be considered at the outset of the operation.

At the tactical phase the aim is to determine precise contours of the anomaly so as to deduce the location and the most likely type of mineralization. The method consists of regular grid sampling, either on a square or rectangular network, or along equidistant sections when the anomaly is known. The sample taken is always a soil sample or where necessary a weathered rock sample **(Fig. 4.5)**.

Location and relocation of sampling points relies on the transferral of the prospecting grid into the field. A compass or theodolite is used to draw a straight baseline at a slight angle (10-20°) to the probable area to be studied. This ensures that the isoanomalies are not arbitrarily placed on the baseline. This baseline is staked out at regular intervals according to the chosen prospecting grid. The two ends of the baseline are marked out on the ground either by concrete markers or by poles different to the others. Perpendicular sections are drawn from each pole to the baseline, and are marked out according to the required spacing. The sampling number is attached to each pole (plastic strip, metal label, etc.). At this tactical stage the relief of a prospected area may play an important part in the interpretation, any displacement of the anomaly by just a few metres being of significance. At each point the direction of the steepest slope will be marked by an arrow. In a mountainous region samples should be taken along the contour lines or a quick topographical survey made. Likewise, if the work is to be accurate, then all the visible phenomena which could be useful in the interpretation of the anomaly must be mapped in detail: outcrops, gossans, quartz veins, ancient mine workings. All of these are shown on the sampling location map, together with the major topographical features of the prospected zone **(Fig. 4.6)**.

It is often sufficient at this stage to locate the sampling points on a rough topographical map (compass, 'topofil'). Later, according to the results and the scope of further work (geophysics, drilling) an accurate topographical survey can be made.

The spacing of samples varies and depends on the problem in question. For example, for an Sb anomaly initial spacing should not exceed 50 or 100m maximum, whereas a grid of 200 to 400m would suffice for the location of Pb–Zn mineralization. The one remaining thing to do is to define the extent of mineralization and the maximum value or 'top' of an average anomaly from a tighter sampling grid.

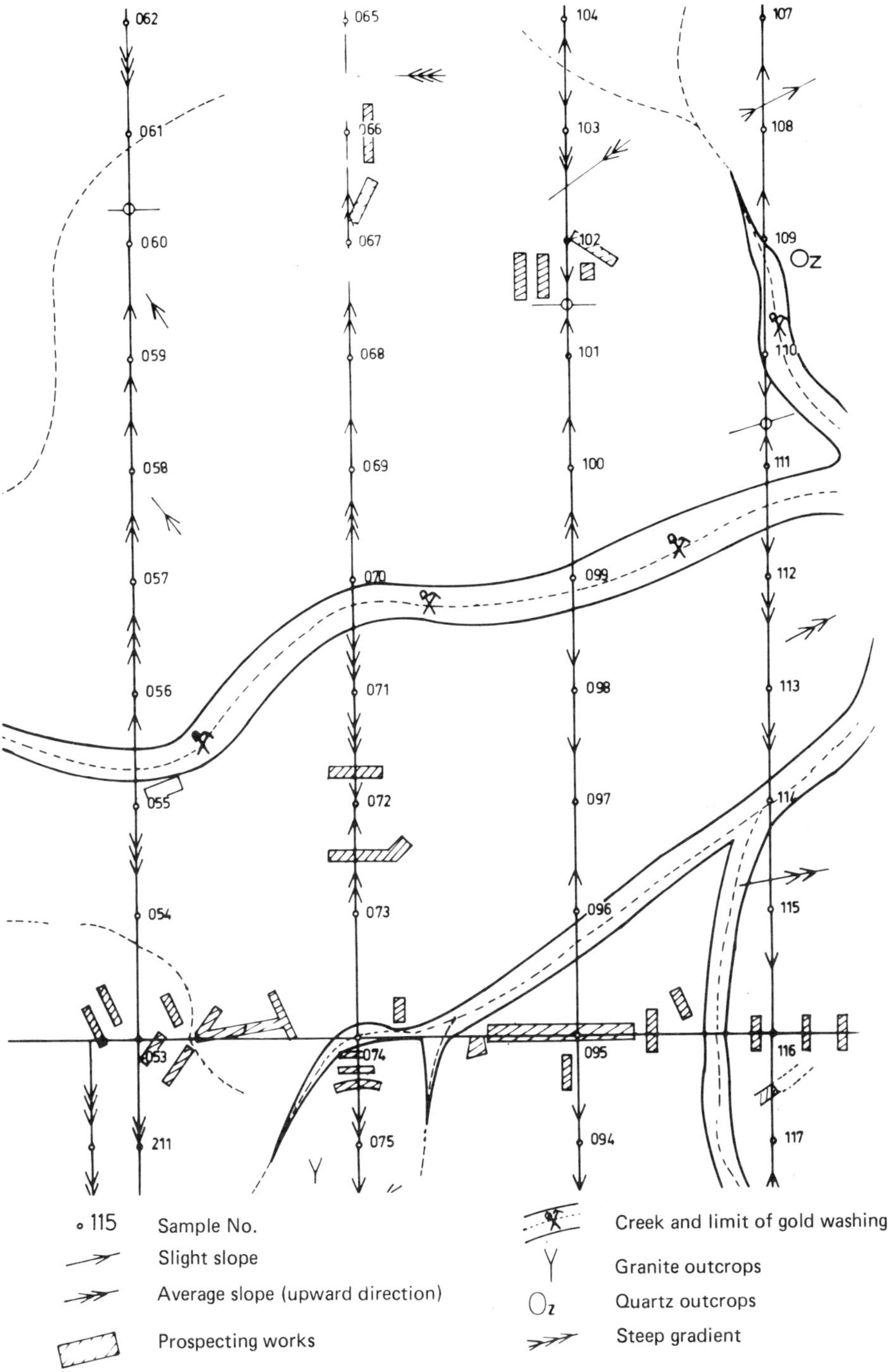

Fig. 4.6. *Type of information to be included on a tactical prospecting map. Grid, 100/50m; scale, 1:2000.*

3.1.1 THE INVESTIGATION OF THE SUB-SURFACE SOURCE OF THE ANOMALIES

Once the highest value anomaly has been defined on the surface sampling at depth can be used to find out if this surface picture corresponds to the area of the geological unit at depth presumed to be responsible for the geochemical anomaly. In order to do this, the anomaly is cut by one or several auger sampling holes. Either hand auger drills ('Helix' type) or motor-driven auger drills (eg Stihl O8S model) are used. The 'Helix' type (Appendix 1, Chapter 3) is a spiral auger, screwed on to the end of round or squared tubes, which is drilled into the ground using circular movements with the aid of a wrench. The ground is usually wetted if it is too dry at the sampling location. This can be done as long as the auger is in level A. However, it should be avoided in levels B and C where samples are often taken, as water moistening introduces foreign ions into the soil. If the bedrock is hard, it is unlikely that it can be sampled using this method.

In this case motor-driven auger drills with an endless spiral screw are used, such as the Stihl O8S model which can drill holes to 20m deep with a diameter of 90 to 250mm. Samples come up to the surface continually but 1, 2 or 3m core sections may also be taken. These drills are capable of cutting a few centimetres into the bedrock and thus chip samples can be taken. Other types of mechanized auger drill exist, which are mentioned in Chapter 7.

The 'funnel' rule of thumb should be referred to whilst on this subject: the deeper the samples, the tighter the sampling grid. The reason for this is that elements are more dispersed in the soil than in the rock which has undergone varying degrees of alteration. For example, to test the bedrock at a depth of 5m below a surface anomaly determined on a 25 × 25m grid rows of 5m spaced auger drill holes should be made **(Fig. 4.7)**. Sampling will vary according to whether the purpose is to study the variations in concentrations between the surface and the bedrock or simply to search for the optimum development at depth.

1) One sample per metre of soil is cut by the auger drill (either by quartering of 1m sample sections or samples constituted just by the last 10cm of the section).
2) The sample is taken solely from the bedrock.

A geochemical auger sample taken in order to find the root of an anomaly may be considered as a rock sample, all the more so because it is close to the bedrock, and at its preparation it should be crushed in the same way as a rock.

If rock chemistry is used for studying primary deposits, sampling is based on core drilling or on cuttings where percussion drilling is employed. Samples are taken every 1 to 5m when drilling through homogeneous formations. In the event of a change in the petrographic facies in the mineralization, each facies will be sampled separately. Each sample will therefore be accompanied by as complete a description as possible.

Along exploratory traverse about 10 chips of a 1 to 2cm^3 are taken at random every 2 to 5m; these chips will be aggregated into one complete sample. Be careful not to take samples solely from a harder more mineralized facies.

Marking is done by sequential numbers plotted either on the drilling or on the platform plans.

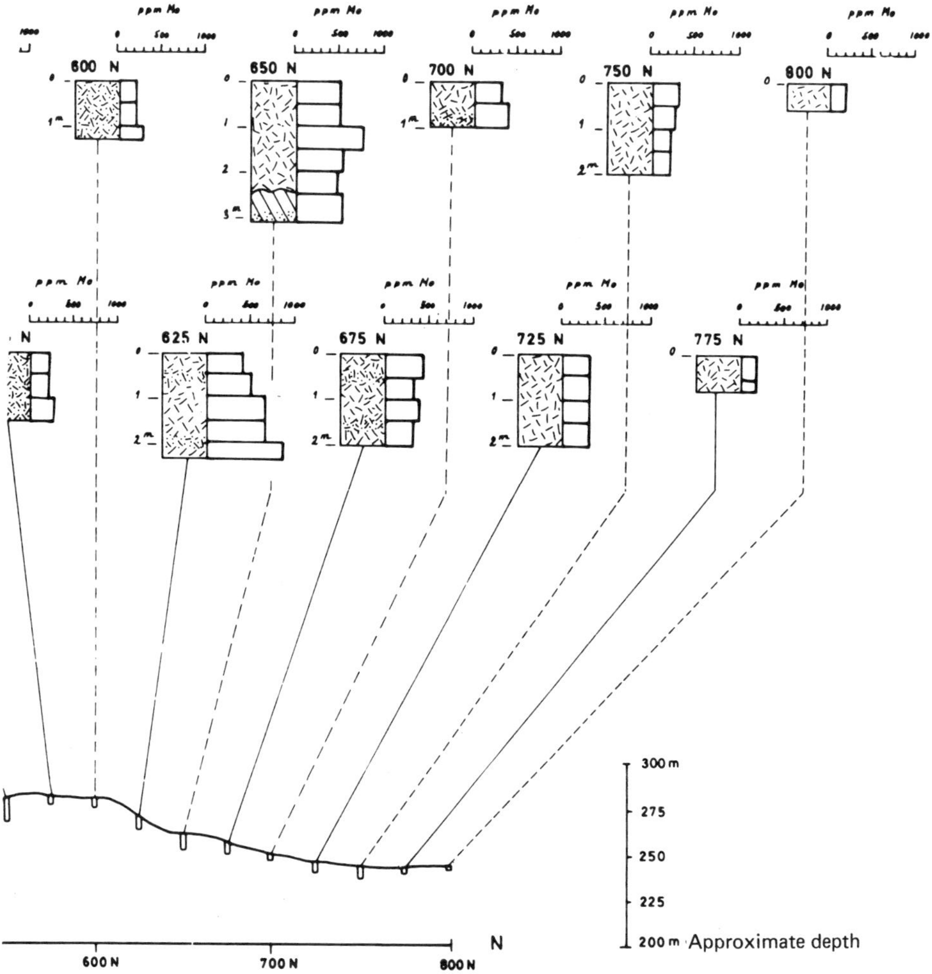

Fig. 4.7. *Augered section below a surface anomaly.*

3.2 Sampling

Once the type of sampling has been fixed and situated on the map the sampler faces the practical problem of knowing where and how samples should be taken in the field. It is impossible to cover every case which may arise. Each situation encountered will be particular but should, in a general and schematic way, fit in with the following possibilities.

3.2.1 SAMPLING IN THE WATER COURSE

The preferred sample is fine grained alluvium from an active bed from beneath the air/water interface but as far as possible away from the current so as to avoid taking overwashed material. This type of sampling is the closest equivalent to the 'stream sediments' idea of English geologists.

Sampling of 'soils' is taken from alluvium or colluvium where the alluvium in the active bed are coarsely grained or absent. This sampling is carried out in the clayey part of the river banks.

Sampling of soils in valley floors is made below level A where there are no characteristic alluvial deposits.

The importance of these water course samples is that they must be to some extent representative of a prospected basin. It is therefore necessary to avoid sampling the main rivers where any mineralized material runs the risk of being diluted by the material of other stagnant water courses. Do not take samples too close to confluences but go sufficiently upstream in the secondary river so as to be sure of taking samples representative of the basin. A sample of white sand is always unsatisfactory even if it is taken away from the current because it does not contain the fine clayey fraction; in the same way you should not sample mud too rich in organic matter (black sample). In tropical countries a good sample is usually anything from grey to brown in colour.

3.2.2 SOIL SAMPLING

Avoid the superficial humic bed. Optimal depth generally varies between 15 and 40cm. The simple technique is to make a mark on the tool used for sampling (sample taken at the depth of a spade, half the length of a machete blade, etc.). Sampling done at the base of herbaceous vegetation roots gives a practical idea of the optimal depth.

In some cases a preliminary methodological study will have determined a special type of sampling, eg surface spreading of sediment in zones. Make sure that the different samplers sample the same thing.

In the usual case the sampling of highly evolved soil (lateritic duricrust or concretions of the calcretic or evaporitic type) should be avoided. Nevertheless it may happen that following a methodological study the consultant geochemist may recommend this type of sampling for a particular reason. Sampling should never be done in strictly allochthonous surface coverings: loess, eolian sand, dune, old terraces of major flats, etc. Auger drill sampling should be used for work in these cover formations.

3.2.3 QUANTITY TO BE SAMPLED

200 to 500g of unsorted material should be taken, according to the proportion of fine sediments included; after seiving, a sample weighing at least 60g should be obtained. A sample weighing about 40g is sent to the laboratory; besides the usual analyses, this quantity is sufficient for checks or analyses of trace elements (Au, Hg, where required, 25g of powder is needed for an Au analysis). The remaining 20g is stored.

In the field the sample is put into a plastic bag measuring approximately 18 × 26cm. A paper bag is not advisable for very wet stream sediments.

The weight of samples to be collected is one of the points to be considered when organizing a return journey. Balance the traverse in such a way that food carrying is replaced by sample carrying; eg in strategic prospecting 10 to 15 samples per day represent 4 to 7kg to be carried per day.

3.3 Numbering samples, field notebook

The prospector should pay particular attention at this stage as an error in numbering can ruin a whole section of the prospecting programme. Two numbering methods have been used.

3.3.1 NUMBER GIVING LOCATION OF THE SAMPLE

In strategic geochemistry number of the main waterway, sub-number of the tributary, right or left, sequential number of the sample, eg 12 R 5, interpreted as drainage course 1, second tributary on the right from upstream.

In tactical geochemistry sequential number of section and distance of sampling from the baseline, eg 150N–60W, interpreted as section 150m North of the point of origin, sample taken 60m West of the baseline. This method is definitely not to be recommended.

3.3.2 SEQUENTIAL NUMBERING

One single number is given as soon as the sample is taken and kept throughout its analysis, processing and data storage. This method is obligatory for computer processing. A pre-numbered field notebook with detachable pages greatly reduces the risk of repeated numbers. This notebook can also be used to record coded or non-coded field observations made while sampling. An example of a field notebook page is given in **Table 4.1**. We recommend sequential numbering for all strategic geochemistry and even for tactical geochemistry (Chapter3).

3.4 Preparation and despatch

When organizing a large strategic prospecting programme it is useful to plan a small sample preparation unit. Normally it is not possible to send thousands of samples each weighing 400g to a central laboratory. The work should be carried out with great care to ensure that the samples are correctly numbered and delivered. This preparation work is divided into three main stages:

1) drying;
2) crushing;
3) sieving.

3.4.1 DRYING

When dealing with stream sediments in moist climates the sample must be dried before sieving. The most reliable method is leaving it to dry in the sun, since this avoids sample contamination and destruction. Lay out a separate drying area equipped with plastic pans for the samples. Do no separate the sample number from the sample during the drying stage. A large number of samples should be dried at

Date: / / **B. R. G. M.** Sampler:

No

Division Zone I No. of 1:50,000 sheet

☐		☐		☐		☐
Stream sediment	0	Uncultivated zone	0	Marsh		Simplified geological code
Soil on alluvitem	1	Cultivated zone	1	Absence	0	
Soil in valley bottom	2			Presence	1	
Soil	3					

OBSERVATIONS*:
Water flow rate: normal, flooding, dry, etc.
Colour of sample: black, brown, red, grey, white, etc.
Cover: sand, clay loess, moraine, etc.
Ferruginous, siliceous zone, etc.
Mine, quarry, embankment, factory, scrapyard, etc.
Vineyard, orchards, crops, meadows, forests, etc.

Soil sampling:
Thickness: stoney soil, normal soil, thick soil
Depth of sampling: ....cm

* Underline as appropriate

(optional):
Latitude
Longitude
Altitude
m

No

No

Alluvial

Table 4.1. *Example of a page from a field notebook.*

the same time in order to keep sample processing a continuous operation (400 samples drying for 100 samples per day processed). Sometimes the sun is not strong enough to dry the samples, so fires must be used. Avoid intense heating (baking of clay, destruction of organic matter, loss of volatile elements Hg, As, etc.). A good idea is to place prospecting pans one on top of the other separated by three or four stones. If there is an electricity supply at the base camp a ventilated cupboard with an infra-red lamp makes a very good oven.

3.4.2 DISAGGREGATION

When the sample is dry the lumps must be carefully disintegrated to restore the sample to its original grain size without crushing the grains. A ceramic or wooden mortar and pestle is used but the latter is not advisable since it is extremely difficult to clean between samples. (Ceramic 30cm diameter mortar and a No 11 pestle.)

The large mineral and rock fragments are removed at the outset to avoid fragmentation. A satisfactory grain size is reached when no coarse grains can be felt between the fingers.

This should not be considered as a grinding stage. Its aim is to restore the sample to its original size distribution to be able to separate homegeneous size fractions for all samples.

3.4.3 SIEVING

In order to separate this homegeneous size fraction the sample is sieved through a mesh of 0.125mm (100 mesh, AFNOR No 22) **(Table 4.2)**. The sieves used should not introduce any foreign materials, they should either be made entirely of stainless steel or else have a stainless steel frame and nylon netting.

AFNOR series				TYLER series			
1	2	3	4	5	6	7	8
Aperture size (mm)	Modules	Volume mm³/10³	Weight density 3,0 gr. 10-⁶	Aperture size	Mesh	Volume mm³/10³	Weight density 3,00 gr. 10-⁶
1 000	31	525	1 574	1 003	16	528	1 585
800	30	268	804	833	20	303	908
				701	24	180	541
630	29	131	393				
				589	28	107	321
500	28	66	197	495	32	64	190
400	27	34	100	417	35	38	114
				351	42	23	68
315	26	16	50				
				295	48	13	40
250	25	8	25	246	60	8	23
200	24	4	12.5	208	65	5	14
				175	80	3	8
160	23	2	6				
				147	100	1.5	5
125	22	1	3	130	115	1	3.5
100	21	0.5	1.5	104	150	0.6	1.8
80	20	0.27	0.8	88	170	0.35	1
				74	200	0.21	0.6
63	19	0.13	0.4	62	230	0.13	0.4
50	18	0.065	0.2	53	270	0.076	0.2
40	17	0.033	0.1	44	325	0.045	0.13
				37	400	0.027	0.08

Table 4.2. *Table of the range of sieves conforming to the AFNOR and TYLOR (USA) norms.*

A 125μm sieve is used since this mesh is very suitable for separating the coarse fraction (quartz, plant fragments) from the fine homogeneous fraction which can then be directly chemically analysed (further grinding is necessary for quantitive analysis). In other than temperate climates experience has shown that this $<$125μm fraction is not always suitable. In tropical climates the clay-fine sand fraction and in desert climates the eolian quartz fraction may be concentrated in this size fraction.

Various methods can be used to try to eliminate the waste and consequently to concentrate the valuable part:

1) Appropriate grain size separation: there is often a slightly coarser fraction.
2) Magnetic separation: this method enables the separation of the resistant quartz-clay fraction, it is carried out in a central laboratory on a Frantz separator.
3) Heavy liquid separation, etc.

The geochemist should decide on a combination of appropriate methods modifying the standard sampling and preparation methods, where necessary. When the metal concerned is known and may occur in the form of flakes which will not pass through the sieve (MoS_2, Au) a coarser size fraction should be used.

After the <125µm material recovered has been divided by the Jones sample-divider or by quartering two large sample tubes are filled (diameter: 3.5cm, height: 7cm). One is sent to the central laboratory, the other is kept at the base or at the local office.

These tubes are numbered by placing round labels on the inside of the stoppers. Special care should be taken at the end of the sieving stage to avoid errors in numbering. The sampling label is kept with the sample throughout the drying and disaggregation stages. At the sieving stage it is put in the pan which collects the sieved material, known as the sieve bottom. Several people can work on disaggregation and sieving with one person receiving the numbered sieve bottoms and filling the tubes.

3.4.4 DESPATCH

See Chapter 3.

3.4.5 SPECIAL PREPARATION FOR ROCK GEOCHEMISTRY: GRINDING

In rock geochemistry or auger bedrock sampling (level C) the objective is to find the exact location of mineralization or its primary deposit. Enough samples should be taken to be representative, they are then quartered and ground to <125µm. The powder can then be analysed.

When the ground material is to be geochemically analysed it must be ground in a non-polluting grinder (agate mortar). As the grinding rarely takes place on site be sure to mark it on the despatch and analysis documents. Non-polluting grinding is very important in rock geochemistry where concentrations are low and the least amount of pollution will be significant.

4 Data processing

4.1 Strategic geochemistry

Strategic geochemistry and multi-element analysis with the numerous data obtained (33 elements/sample) requires the data to be processed on a computer. This processing normally takes place at the technical centre, as is the case with the BRGM, thus the prospector has only to provide location maps and field data (sampling notebook).

In Appendix 1 the sequence of possible procedures is presented schematically and a brief explanation is given of the various methods of statistical analysis used.

Presentation of the results is based on computer-aided mapping software developed by the central geochemistry unit. This includes:

1) maps of raw data;
2) maps of anomalous values using different patterns;
3) maps of local geochemical backgrounds.

These documents enable an overall map to be drawn up showing the anomalous areas which warrant further investigation.

4.2 Tactical geochemistry

The data is less abundant and the systematic use of a computer is not necessary. Consequently, the prospector often has to process and present the data himself.

An elementary statistical analysis aimed at separating background from anomalous contents can be done by hand (calculation of means and deviations).

There are two types of distribution parameters:

1) Central tendency characteristics representing the order of magnitude of the concentration in one element of all samples. These are:

The arithmetic mean representing the quotient of the sum of the concentration of one element by the number of samples:

$$\bar{x} = \frac{1}{n} \sum_{i=1}^{n} xi ; \ xi = \text{ concentration of sample i}$$

The geometric mean G is obtained by averaging the grades transformed into logarithmic values:

$$\log G = \frac{1}{n} \sum_{i=1}^{n} \log xi.$$

This mean calculation usually gives a better estimation of the geochemical 'background concentration' than the arithmetic mean by decreasing the relative importance of the extremely anomalous values.

2) Dispersion characteristics show the extent to which the various observations of one series differ from each other and thus also from the chosen central value.

Variation interval is the difference between the extreme values of the variable studied:

$$I = x_{max} - x_{min}$$

Standard deviation is a dispersion parameter which takes into account how the observed values deviate from the mean:

$$\sigma = \sqrt{\frac{1}{n-1} \sum (x_i - \bar{x})^2}$$

Coefficient of variation is defined as the ratio of standard deviation to the arithmetic mean:

$$V = \frac{\sigma}{\bar{x}}$$

This coefficient enables comparison of the dispersion of both series. It is a nondimensional parameter which does not take into account the order of magnitude of the variables.

Geometric deviation (geometrical error ε) is the number whose logarithm corresponds to the standard deviation of the logarithmic values of the concentration:

$$\log \varepsilon = \sqrt{\frac{1}{n-1}\sum (\log x_i - \log G)^2}$$

The distribution parameters are normally used to estimate the anomaly threshold values (A) for the elements studied. Thus, for example, the threshold value which enables 2.5% of the higher values (anomalous values) to be identified is calculated as:

- for a normal distribution:

$$A = \bar{x} + 2\,\sigma$$

-for a log-normal distribution:

$$A' = G\varepsilon^2$$

Histograms can be drawn without difficulty: divide the scale of x into about 20 regular intervals (10 in 10ppm, 50 in 50ppm, according to the elements). At each interval draw a rectangle with an area proportional to the observation frequency within this interval. The general shape of the enveloping curve obtained indicates whether the distribution is regular or not and enables selection of representative cut-off values.

Cut-offs determined only by statistical methods are independent of various natural phenomena which have to be taken into consideration.

If the prospecting involves a geologically varied terrain, this can cause significant variations in certain elements thus creating formational anomalies which may conceal true anomalies: for example, certain basic rocks contain 10 to 20 times more copper than granite rocks; Zn is concentrated in a carbonate environment, whereas it is often leached in an acid environment. Thus it is necessary to determine as far as possible the goechemical and the anomaly thresholds separately for each geological domain. The cut-off values having been determined, the results are normally presented using isocontent maps. Contour spacing should be selected in order to show anomalous zones and the gradients in marginal zones. The colours used should where possible show a gradual progression from low to high grade.

Low grade white
light brown
brown
dark brown

High grade bright orange

Charts which have a distinct colour at both ends of the scale (eg red and blue) should be avoided as there is a tendency to focus on one point on the map.

Do not use so-called synthesis maps where several elements are plotted as they prevent an understanding of the geochemical environment. However the main details must be given (with reserve). These include any old workings known as well as mineral showings. The same applies for topography.

The best method for studying variations in several elements is to superimpose single-element maps using tracing paper.

5 Interpretation and selection of anomalies

The aim of all geochemical prospecting is the identification of anomalies, necessitating the collaboration of prospector and geochemist. The prospector provides field information, descriptions of showings encountered, information on possible pollution, etc. The geochemist uses his knowledge of differential element mobility, associations characteristic of different formations, etc. He will also be able to refer to previous examples he has worked on.

Appendix 1

Geochemical data processing

1 Introduction

Multi-element geochemical prospecting from initial sampling to the presentation of results involves a series of diverse operations which can be presented schematically as a flow chart **(Table 1)**.

The first stage involves the acquisition of geochemical information from samples collected in the field. This is achieved by regrouping in the same index file:

1) analytical results supplied on paper tape by the laboratory;
2) coordinates of the sample locations with the help of a coordinating graph (measuring protractor);
3) field observations (geology, type of sample) transcribed after coding, with the aid of perforated computer cards.

This index file is then recorded and stored on magnetic computer tape or disk. The results from then on are consequently available for any data processing such as editing, statistical calculations, cartographic plotting, etc.

Data processing essentially involves the mainpulation of a large number of variables (the sample values). It specifically determines the variability within and between the sample populations.

Table. 1. *Multi-element geochemical prospecting.*

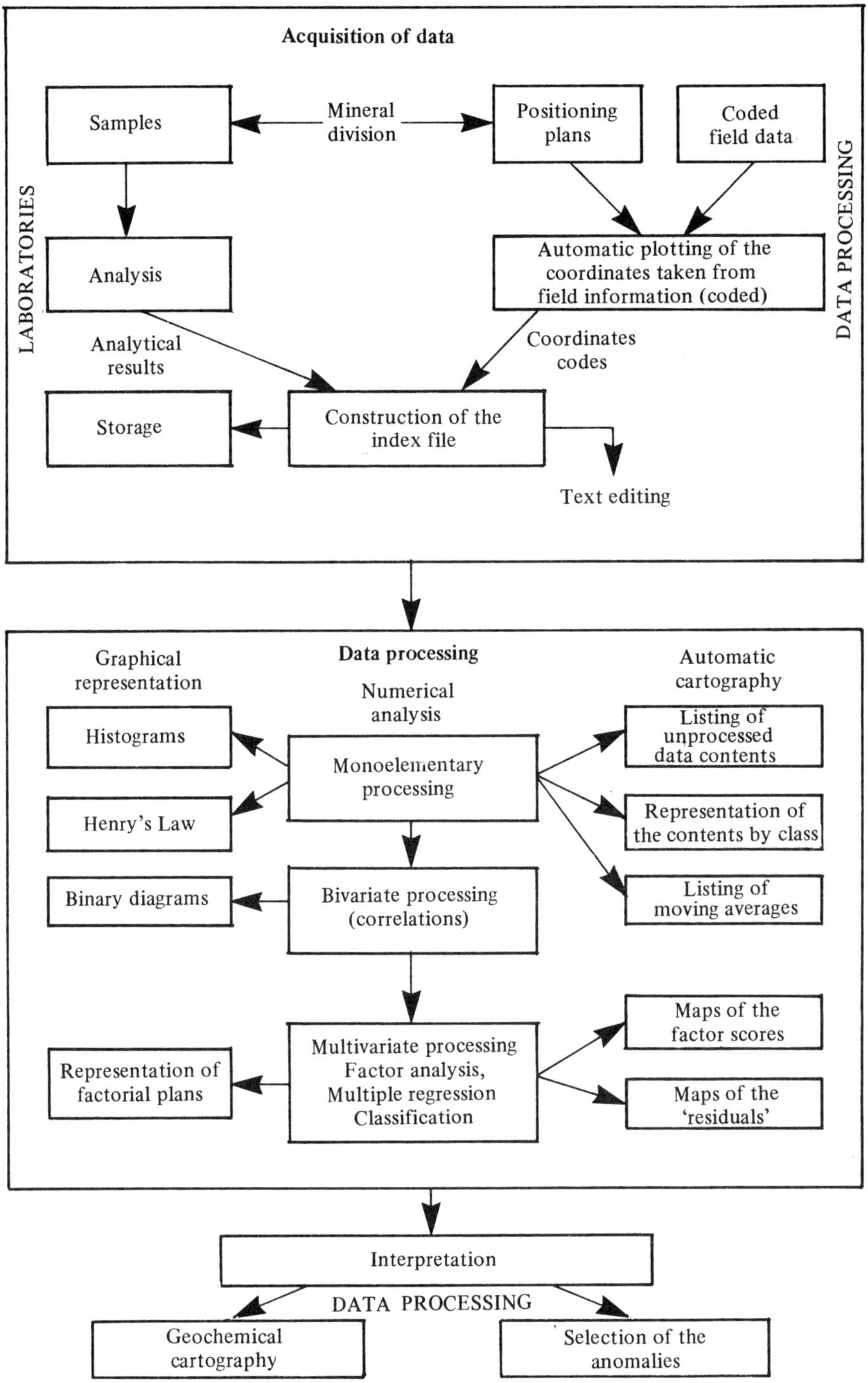

Three large groups of statistical methods are used: the first involves the processing of variables, taken one by one (univariate analysis), the second groups the study techniques of the concomitant variations of two characters (bivariate analysis) and the third processes the data simultaneously (multi-variate analysis).

UNIVARIATE ANALYSIS

Univariate analysis or elementary analysis allows the essential characteristics for an element distribution to be summarized either by calculation or graphically. The characteristic parameters for the distribution which have been defined for their use by the prospector in Section 4 are calculated directly by computer and grouped into a table.

The graphical representation for the given element distribution allows one to determine which statistical law best fits the distribution law for the contents or if there exists differentiate populations within the global sample obeying the different laws.

STATISTICAL BIVARIATE ANALYSIS

Statistical bivariate analysis consists of analysing two character simultaneous variations, either on a graph or by calculation of the linear correlation coefficient.

STATISTICAL MULTIVARIATE PROCESSING

Multiple regression

Multiple regression, followed by a calculation of the remainders, allows variations of a variable (known as dependant variables) to be linked to variations of one or several other variables (known as explicative variables). It aims essentially to give prominence to and to eliminate the metallic materials from the primary deposit. Thus for example, the high levels of copper associated with basic rocks could be suppressed, even deleted by the study of the distribution of Ni, Co, V. On the other hand, the really significant anomalous values will persist and appear better contrasted.

Factorial analysis (A Herboch and P Laffitte)

The objective of factorial analysis is to extract, under a condensed form, the largest item of information contained in a large volume of numerical data (in this case the results of geochemical analysis). It involves explaining the relations observed between the fluctuations of the levels of various measured elements, in terms of the simplest relations and above all the most basic. This synthesis requires a series of complex mathematical calculations, which have as their starting point the table of correlation coefficients between the variables.

It results in the creation of a small number of hypothetical variables (factors) taking into account the largest section of observed variations. These factors, in favourable cases, will reflect the phenomena which form the origin of these variations; in addition, one will try to identify each of them with a reality (geological, mineralogical, etc.), that is to say, to interpret the results for the calculations in a 'true' perspective.

If for example a plutonic series is studied, beginning with chemical data and Fe, Mg and Ti are grouped on the same factor, this could express the variation in the levels of ferromagnesian minerals in the different samples. In the case of geochemical prospecting the facts will be able to illustrate the presence of various mineralization, the contrast between the main geological units, the pedological phenomena, etc.

The influence of each of the factors in a particular sample can be evaluated using a figure called a 'score' of the factor for the sample being considered. Cartographic representation of these factor 'scores' then allows the importance of the phenomena at the correct scale for the region illustrated by the factor in question to be visualized.

CLASSIFICATION OF SAMPLES

Classification of samples employs direct or analogical techniques. They allow regrouping of the individual samples presenting neighbouring geochemical spectra and thus contribute to distinguish the characteristic units of the lithological or metallogenic entities. These techniques often profitably complete the information provided by factorial analysis, for cartography as well as for research of anomalies.

AUTOMATIC CARTOGRAPHY OF DATA USING A GRAPH PLOTTER

The various data (grades, factor scores derived by factorial analysis, residual grades) can be mapped in different ways:

1) listing the unprocessed data of the different variables;
2) listing the values, by using different figures for the chosen limits;
3) calculation and listing of moving averages: this method of smoothing consists of dividing the sector being studied into individual cells and calculating for each one of them the average grade of a given element. This technique enables the regional variations of the distribution of element concentrations to be defined. Generally, the cells are superimposed upon one another by half, both in the vertical and horizontal direction; thus each sample is put into four cells;
4) construction of contours of equal concentrations;
5) representation of the fluctuations of the smoothed values with respect to a variable by automatic drawing of a regular framework of points whose size is proportional to the value taken by the variable.

Chapter 5

Geophysical prospecting

In this manual intended for the geologists and non-geophysical prospectors, there is not enough room to cover all the geophysical methods. Therefore, we provide some insights into the subject and the technical advice that could help implement simple methods using non-complex equipment.

When a geologist or prospector is faced with a mineralized showing (lack of continuity or outcrops) far from any geophysical service that could advise him or help him with some tests it would be very useful to have a few simple inexpensive instruments at hand and to know how to operate them. It is not profitable for geophysicists to travel long distances to carry out a few tests. This is why we provide information here to enable geologists and prospectors to apply simple geophysical methods themselves (magnetism, self-polarization, VLF, electromagnetism, scintillometry), together with a brief description of the equipment, while warning against problems which are likely to occur and the pitfalls of interpretation.

A simple geophysical method does not imply that the method is easy to use. Reading is easy with modern push-button instruments, but it is often enough to overlook just one precaution for the measurements to be completely distorted, and the poor result obtained can affect the future of the prospect.

How many geophysical surveys have been conducted with an improper measurement grid or a poor profile orientation? How many magnetic maps have been surveyed with a badly adjusted magnetometer? How many self-polarization surveys have been produced with inadequate wetting of the holes or even with electrodes which were not impolarizable? How many VLF profiles are too widely spaced or badly oriented? How many earthing configurations have been recorded on a pocket handkerchief? Too often the scientist is unaware of his own ignorance and blames geophysics for his failures.

This chapter is not only intended to familiarize isolated geologists and prospectors with the pitfalls hidden under the seeming ease of certain geophysical methods and the interpretation of results, but chiefly to reduce costs and errors.

1 Overall view of geophysical methods

Everyone knows that geophysics, by measurements taken above ground (and sometimes in drill holes), can provide a picture of what is happening underground. Measuring physical parameters helps to locate deposits for structures which contrast (by density, magnetism, conductivity, etc.) with the surrounding material.

Table 5.1. *Geophysical methods applied to mining prospecting.*

Areas of application		METHODS		Property	Parameter measured or calculated	Unit	Direct prospecting	Prospecting: indirect	Prospecting: structural
Non directional methods	RADIO METRIC	Scintillometry		Gamma radiation	Gamma radiation	Shock/second (s/s)	Ores of Uranium, of Thorium, of Phosphate	Clays ≠ granites	X
		Spectrometry					Monazite		X
		Gravimetry		density	acceleration of gravity	milligal (mgal)	Massive sulphides – Magnetite – Chromite – Barite – Coal	Cupolas	X
		Magnetism		Magnetic susceptibility	Magnetic field	gamma (γ) nanotesla (nT)	Magnetite – Pyrrhotite – Chromite – Ilmenite		X
		Self potential (S.P.)		conductivity and redox potential	potential difference	millivolt (mV)	Massive sulphides – graphite		X
Directional methods				CONDUCTIVITY			Prolongation of **productive mineralizations**		·
	ELECTRIC	Resistivity	Sections and maps		apparent resistivity	ohm-metre (Ω m)	Ores | conductive / resistant | and lodes	Faults	X
			Electric "drilling"				Roofs of ore bodies | conductive / resistant		X
		Induced **Polarization (I.P.)**	time-domain frequency-domain	chargeability (+ resistivity) **polarizability**		millisecond (ms) %	Disseminated sulphides Clays – Bauxite Chromite	Faults	
	ELECTRO-MAGNETIC	Electromagnetism	Sections and maps	CONDUCTIVITY	Many parameters: amplitude and phase of field – angle of tilt – lengthening of ellipse of polarization. Components in phase and in quadrature **résistivite apparente**		Conductive ores and lodes	Faults	X
			Frequency-domain						X
		Magnetotelluric	Sections and maps				Conductive ores and lodes	Faults	X
		Seismic	Refraction	elasticity	velocity of waves	second and metre/second (m/s)	Salt domes	Structures	X
			Reflection						X

Less well known are the fact that each method generally involves only one physical parameter, that geometrical factors are very important and that it is unwise to detect a deposit or structure using just one method.

Table 5.1 gives an overall view of geophysical methods and the mining objectives to which they can be applied. If the geologist or prospector knows what he is looking for, and if he knows the physical properties of the ores he is trying to locate, the formations whose structure he is trying to decipher, then Table 5.1 will enable him to choose the correct geophysical method to solve his problem.

2 Self-polarization

2.1 Introduction

Between two points on the ground at any given moment a potential difference always exists, generally small, ranging from a few millivolts (mV) to several hundred milivolts. These potential differences occur throughout the frequency spectrum. Here we are not interested in rapid variations (1 to 1000Hz) from industrial (stray currents, especially 50Hz) or natural (telluric currents) sources and those that are generated by certain geophysical methods (telluric, magnetotelluric), but we only consider d.c. voltages or very low frequencies generated by:

1) electrochemical phenomena (cell effects);
2) electrokinetic phenomena (filtration effect).

The electrochemical phenomena generated by sulphides are used in mining geophysics to detect these sulphides in the ground. Electrofiltration, together with electrochemical polarization of graphite, often distorts the effects caused by sulphides.

2.2 Principle of the method

2.2.1 CAUSE OF THE PHENOMENON

When an electrode with metallic conductivity, ie the current is produced by the movement of electrons (in the case of metals, most metallic sulphides and graphite) as opposed to ionic conductivity in which the current is created by the movements of ions (in the case of rocks containing water), is immersed in a solution a potential difference is created between the electrode and the solution. This is termed the electrochemical mechanism of electrode polarization. If the solution has a variable composition and pH, an electrical voltage is created across the terminals of the conductor. This is termed the cell effect.

If a deposit of sulphides (such as pyrite) is located partly above the water-table, a cell is created because the deposit behaves like a body with metallic conductivity, of which the top is in an oxidizing medium and the bottom is in a reducing medium: the differences in redox potential and electrolyte composition cause a negative pole to appear at the top of the deposit and consequently negative potentials at the surface directly above the deposit. The potentials are measured to detect the nature of the deoposit **(Fig. 5.1)**.

Minerals which generate SP effects include pyrite, pyrrhotite, chalcopyrite, bornite, mispickel, covellite, chalcocite, galena and graphite. Sphalerite, which is not a conductor, does not produce SP.

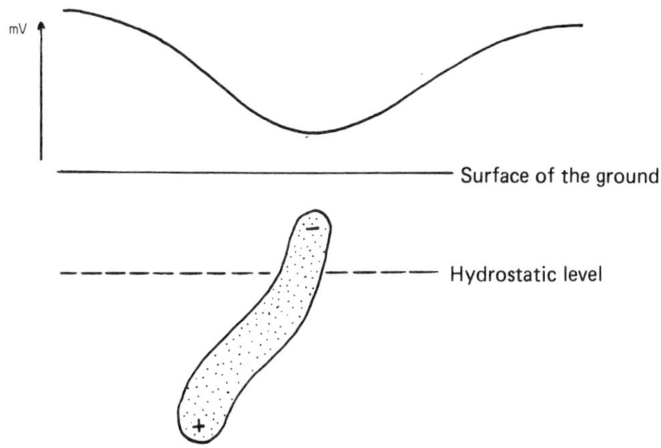

Fig. 5.1. *SP anomaly caused by a pyrite lens.*

2.2.2 MEASUREMENT PRINCIPLE

The measurement of natural ground potentials requires:

1) a d.c. millivoltmeter with high input impedance (because the earth connections are highly resistive);
2) impolarizable electrodes (or the polarization of the electrodes in contact with the ground will create an interference potential and mask the desired effect);
3) a cable connecting the electrodes to the instrument.

2.3 Equipment and measurement

2.3.1 MILLIVOLTMETER

Use was formerly made of potentiometers in which the voltage across two electrodes was measured by a contrast method. Electronic mvs with pointer scales are now nearly always used (BRGM SP meter, Philips millivoltmeter), together with the very latest digital display.

We describe the BRGM SP here as an example. Beginning at the top of the meter it can be divided into the following parts:

1) two electrode connecting terminals,

2) pointer scale and below it mechanical zero adjustment knob;
3) switch (left) and battery check (right);
4) electronic zero adjustment (right);
5) range selector.

This instrument is very easy to use:

1) adjust the instrument at the beginning of each day;
2) check the batteries: this must be performed when the instrument is connected and running;
3) check mechanical zero: when the instrument is 'off' bring the pointer to zero by turning the knob located just below the scale;
4) check electronic zero: when the instrument is 'on' bring the pointer to zero by turning the other small adjustment knob;
5) connect the electrodes by linking them to the terminals of the instrument (mobile electrode at M, reference electrode at N: measure V_M–V_N);
6) check that the range selector reads 100 or 300mV (to prevent the needle from shooting up to the end of the scale);
7) set the switch to 'on' and take the reading, making sure to:

   - record the sign of the deviation (+ or –);
   - take the reading on the correct scale (upper scale: 100 divisions correspond to 1000mV on scale 1000 for example; low scale: 30 divisions correspond to 300mV on scale 300 for example).

The meter measures 20 × 15 × 13cm and weighs 2kg (with six 1.5V batteries).

Precautions to be taken:

1) shelter the instrument from rain: if certain components get wet, the pointer goes haywire and the measurements must not be taken until the instrument has been dried out;
2) avoid collisions: the instrument is robust but collision can damage the moving parts of the galvanometer;
3) remove the batteries when storing the instrument because they are liable to leak and damage the electronics or at least the contacts of the battery case.

2.3.2 IMPOLARIZABLE ELECTRODES

It is impossible to measure the SP using metallic electrodes (steel, copper, lead) inserted into the ground. This is because the redox potential of the ground is not the same at each point, and electrode polarization creates an interference voltage, which is often larger than the potential to be measured. This is why it is necessary to use so-called 'impolarizable' electrodes, in which electrode polarization is controlled.

We now describe the copper-copper sulphate (Cu–$CuSO_4$) electrode which is routinely used in France **(Fig. 5.2)**.

A porcelain beaker with a porous bottom is filled with a saturated solution of Cu–$CuSO_4$. A copper rod connected to the SP meter by a cable is immersed in the solution. Since both electrodes are absolutely identical, the potential created at the contact of the rod with the solution will be the same (within 1 to 2mV) if the electrodes are balanced. Thus the contact between the electrode and the ground occurs without any significant interference voltage.

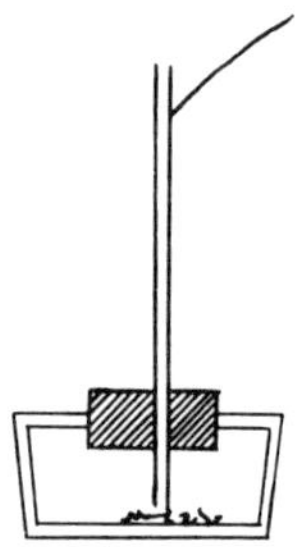

Fig. 5.2. *Impolarizable electrode.*

Other types of impolarizable electrodes are available (eg cadmium, cadmium sulphide). The porcelain beaker can be replaced by a porous wooden box.

2.3.2.1 Preparing the holes to position the electrodes

Contact between the electrodes and the ground must be good. This means that plants, grass and roots must be removed from the place where the electrode is to be inserted. If the ground is wet, one blow of a pickaxe is enough to remove the grass and then shape the hole with your heel just before inserting the electrode and taking the measurement. If the ground is dry, the hole must be wetted with ½ litre of water in advance so that the water soaks evenly through the ground by the time the measurement is taken. If the water is added too early the hole may dry out before the measurement is taken. This is crucial in the success of measurements in hot and dry countries.

Avoid positioning SP stations near pipelines, earth connections, power pylons, transformers or even trees, so that interference voltages are eliminated.

When proceeding on bare rock a copper rod wrapped in a sponge soaked in a saturated solution of $CuSO_4$ can be used as the electrode.

2.3.2.2 Precautions

These electrodes are fragile. Avoid collisions because any crack causes the electrolyte to leak, which is liable to unbalance the electrodes. The electrodes must be prepared in advance before the survey begins to allow them to reach equilibrium.

From time to time (twice a day) check the balance of the electrodes. To do this, place both electrodes in a plastic bowl containing a little water (or $CuSO_4$ solution), connect them to the SP meter, take the reading and record it. Two well-balanced electrodes should not differ by more than a few millivolts. If the discrepancy exceeds 10mV, drain the electrolyte from both electrodes, homogenize it, refill the electrodes and allow to stabilize for about 1h.

Do not leave an electrode fixed to the ground in open sunlight while moving the other electrode around. A difference of 10°C between the electrodes creates an imbalance of about 1mV.

Take care to avoid mistakes when connecting the electrodes. An inversion changes the sign of the measurement between two points.

2.3.3 CONNECTING THE CABLE

The electrodes must be connected to the SP meter by an insulated cable. This cable may be narrow since it carries a very small current, but it must exhibit good tensile strength. If the fixed electrode method is employed, about 400 to 600m of cable are required, wound on to a light-weight reel. Crocodile clips can be used for connections.

2.3.3.1 Precautions

The insulation of the cable and the connections must be perfect. Any metal part in contact with the ground (even if only through the moisture of the cable) will create serious interference. It is especially important to check that the cable of the 'topofil' does not cut through the cable insulation.

This insulation can be checked by immersing the coil in a container filled with water (allowing the connector to project above the surface). The insulation of the coil can also be checked by taking a measurement with the coil placed on the ground and another with the coil insulated from the ground.

2.4 The measuring system

Measurements are always taken with two electrodes M and N but three methods are available.

2.4.1 FIRST METHOD

A reference electrode N is installed permanently in a corner of the survey zone. If, by accident, it is placed on an SP anomaly, this is observed from the very section and it is preferable to change its position. The other electrode M is mobile and is placed successively at the measurement stations.

It is simplest for the operator and his SP meter to travel with the mobile electrode. The cable from the reel is then connected to the fixed electrode. The operator can also remain stationary and have the mobile electrode (and the reel) moved by an assistant, but this is not recommended because a means of communication (eg walkie-talkie, whistle, trumpet, horn) between the operator and his assistant is required, with all the drawbacks that this entails: additional personnel, misunderstandings, uncertainty over the markings, etc.

The station at which the fixed electrode is located is taken as the origin of the potentials. More precisely, it is assigned that value read at the time of checking the balancing of the electrodes.

2.4.2 SECOND METHOD

This employs one electrode fixed at the intersection of the section with the baseline. The same procedure is followed, but it is also necessary to make an SP survey of the baseline and to take into account the potential at these intersection points to calculate the SP values at each station (whereas with the first method, the measurements could be transferred as they were taken, assuming the electrodes balanced).

2.4.3 THIRD METHOD

This method uses a short cable (equal in length to the distance between stations) and both electrodes can be moved at each measurement. This method offers the advantage of not requiring the release (and rewinding) of the cable, but it also has the disadvantages of being more prone to accidental errors and to electrode imbalances, and measurements between the point selected as the origin and the station must be added together to find the potential of the station. Two procedures can be followed.

2.4.3.1 First procedure

If a is the imbalance between two electrodes and V_1 the potential at the first station, it can be seen that the potential at the fifth station will be:

$$V_5 = \Delta V_1 + \Delta V_2 + \Delta V_3 + \Delta V_4 + 4a + V_1$$

in other words this imbalance is cumulative, and this is a disturbing fact if the traverses or circuits are long **(Fig. 5.3)**.

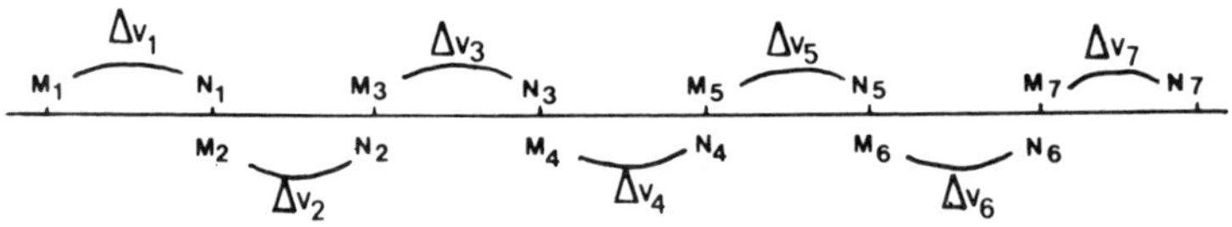

Fig. 5.3. *First procedure.*

2.4.3.2 Second procedure

In this case, the imbalance is compensated, but the risk of error in sign is high **(Fig. 5.4)**. In the case of an accidental error let us assume that the operator makes an error in taking the reading at one station only:

Fig. 5.4. *Second procedure.*

1) using the first method only this station is affected by the error which passes undetected, unless the difference to the neighbouring stations appears suspect;

2) using the second method the same thing occurs, unless the error is made during the survey of the baseline, affecting a complete section;
3) using the third method all the stations (of a traverse or of a circuit) surveyed after the incorrect reading are affected by the error. This is why measurement circuits are generally practised and the correct (or incorrect) 'closure' helps to guarantee the absence (or presence) of one or more errors.

Whenever possible, we therefore recommend using the first method (fixed electrode), which is safer and offers a maximum of advantages.

Dimensions of a survey: blocks over 20 or 30ha in area should not be surveyed. For rows 300 to 400m long the d.c. component of the telluric currents disturbs the measurements and the corresponding gradient is observed on the map. If the survey zone is too extensive, it is preferable to survey adjacent blocks and to connect them by measuring common sections. If the SP gradient is too pronounced, it must be corrected. These gradients are generally 10 to 20mV/km but a gradient of 100mV/km has already been observed in the Poitiers trough, for example.

2.5 Network of stations

SP is a potential method. Each station must be surveyed independently of the others and placed at any location. Consequently, we recommend taking measurements on sections already plotted for other methods (geochemistry, magnetism, etc.) or on the identified rows of a topographical map (boundaries or diagonals of plots on the cadastral survey).

Only in the absence of landmarks (eg forest) is it necessary to draw the sections and leave markers at certain stations, to return there if required. As in all methods, the sections must be plotted perpendicular to the known or presumed structures and, if an anomaly is identified, an extra section must be plotted more or less perpendicular to the axis of the anomaly.

A suitable spacing for measurements is 20 to 25m (with extra stations in the case of a high gradient). Very little is to be gained by a wider spacing, given the short duration of the measurement and the transfer time between stations.

The spacing of the sections must be between 50 and 100m. To obtain a valid SP map the measurement density must be between 5 and 10 points per hectare. The average output is 100 to 200 stations daily.

A good location of sections and stations is an essential factor for success, as for other geophysical methods. It helps to draw a suitable map and, above all, to return to the field subsequently for other operations that may be required **(Table 5.2)**.

Table 5.2. *Estimation of the depth of the top of the deposit.*

Horizontal extension	d approaches
0 (cigar)	From the half-width of the anomaly to 2/3 of its amplitude
2d (vein)	From the half-width of the anomaly to 3/4 of its amplitude
6d (vein)	From a quarter of the width of the anomalyy to 3/5 of its amplitude
10d (vein)	From a quarter of the width of the anomaly to 2/3 of its amplitude

2.6 *Interpretation of SP and its pitfalls*

2.6.1 AMPLITUDE OF SP ANOMALIES

This amplitude depends on too many parameters to guide interpretation. The only empirical rules that can be given are the following:

1) if SP yields anomalies exceeding 500mV, it is highly probable that graphite is present;
2) pyrite lenses of economic size generate anomalies from 100 to 200mV as a rule but some may be as high as 400 to 500mV;
3) networks of sulphide veinlets generate anomalies under 100mV.

2.6.2 NATURAL INTERFERENCE

Electrofiltration of the water descending through porous formations (sands, sandstone) may generate significant negative interference possibly exceeding 1V. SP isoanomalous rows must always be compared with contours. If the correlation is too close, electrofiltration can be suspected.

Electrofiltration of ascending water is encountered in hot countries with high evaporation. Positive potentials of about 10mV are therefore found on sandy desert areas.

Note that sometimes high negative potentials (about 100mV) are observed at stations too close to trees, and temporary effects are caused by passing overhead electrically charged clouds.

2.6.3 ESTIMATION OF THE DEPTH OF THE TOP OF THE DEPOSIT

The depth (d) of the top of the deposit yielding the SP can be estimated from the shape of the sections across the anomaly. The case below refers to steep deposits ('cigar' or vein) with a wide extension at depth.

2.7 *The field log: corrections*

We advise recording the measurements on a pad whose sheets can be detached as the sections are measured and stored safely in the operator's bag **(Table 5.3).**

As long as the imbalance between electrodes does not exceed 5mV, it is not necessry to make corrections. If the difference is greater than 5mV, it is preferable to replace the electrodes, and if possible to make frequent checks to measure the imbalance and correct measurements.

The sections and stations must be recorded on a summary map during the survey.

2.8 *Example of SP prospecting*

2.8.1 ANOMALY DETECTED WEST OF BODENNEC

This example **(Fig. 5.5)** is instructive because it shows how an anomaly, detected by an E W traverse from the road, was plotted roughly by two other traverses in a star pattern. Naturally, this anomaly was then defined with greater accuracy, especially its western tip.

Table 5.3. *A record sheet.*

Survey: VOSGES 22 June

Section M

Reference electrode: Base B3

	Balance check	
	Reference electrode N;	mobile electrode M
	8.00 a.m.	+3mV
	12.30 p.m.	-1mV

STATION	READING	OBSERVATIONS
0	+21	Intersection of baseline with section M
20	+43	
40	+30	Plot boundary-fence
...	...	....
...	...	
...	...	

It is due to stratas of massive sulphides dipping NNE and approaching to within a 10 to 20m of the ground surface directly below the centre of the anomaly. It is here that the wagon drill holes intersected the massive sulphides for the first time.

2.8.2 ANOMALIES DETECTED AT LA PORTE AUX MOINES

This SP survey was conducted on a lead-zinc geochemical anomaly **(Fig. 5.6)**. The standard 50 × 20m grid helped to identify an SP anomaly tested by drill hole PAM 1. A lens of massive sulphides (pyrite, sphalerite, galena, chalcopyrite) was intersected at a depth of 25m.

However, the -150mV anomaly visible SE of the survey does not correspond to sulphides but to graphite shale. The operator should normally have continued the survey on scarps of Saint-Michel in order to circumscribe the anomaly.

3 Magnetism

3.1 Principle of the method

If any substance (a rock or ore, etc.) is subjected to an inductive magnetic field (eg the earth's magnetic field), this substance is magnetized by influence and the induced field is superimposed on the inductive field. The ability of this substance to be magnetized is called the magnetic susceptibility.

If the induced field acts in the same direction as the inductive field, the substance is said to be paramagnetic; this is the case for common ores and rocks. If the induced field acts in the opposite sense to that of the inductive field, the substance is said to be diamagnetic; this is the case for water, salt and anhydrite.

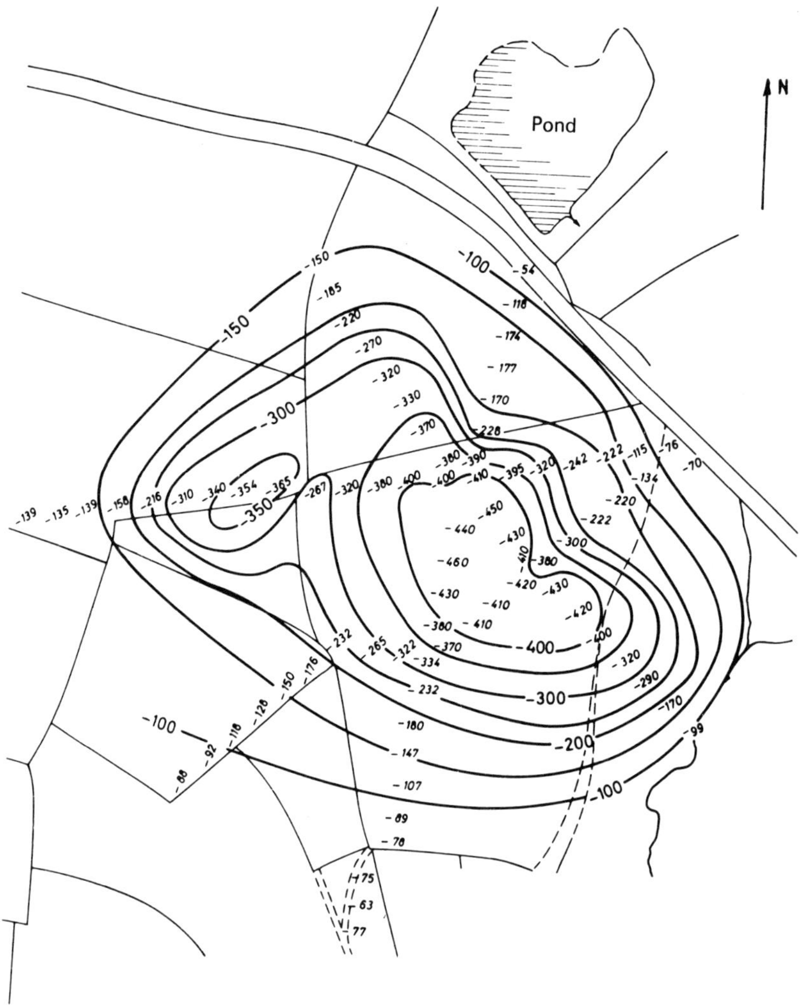

Fig. 5.5. *SP survey on anomaly W (simultaneous to earthing); scale, 1:2500.*

Some substances (such as magnetite) display magnetization in the absence of any inductive field. This remnant field does not necessarily have the same direction as the earth's magnetic field. These ferromagnetic substances, which exist in varying amounts in rocks, impart to them a certain remnant magnetization.

Magnetic susceptibility of main magnetic minerals (in 10^{-6} uem CGS) **Fig. 5.7).**

Magnetite	300,000	to	800,000
Pyrrhotite	5,000	to	30,000
Ilmenite	30,000	to	45,000
Chromite (always contains some magnetite)	3,000	to	50,000
Wolframite	1,000	to	3,000

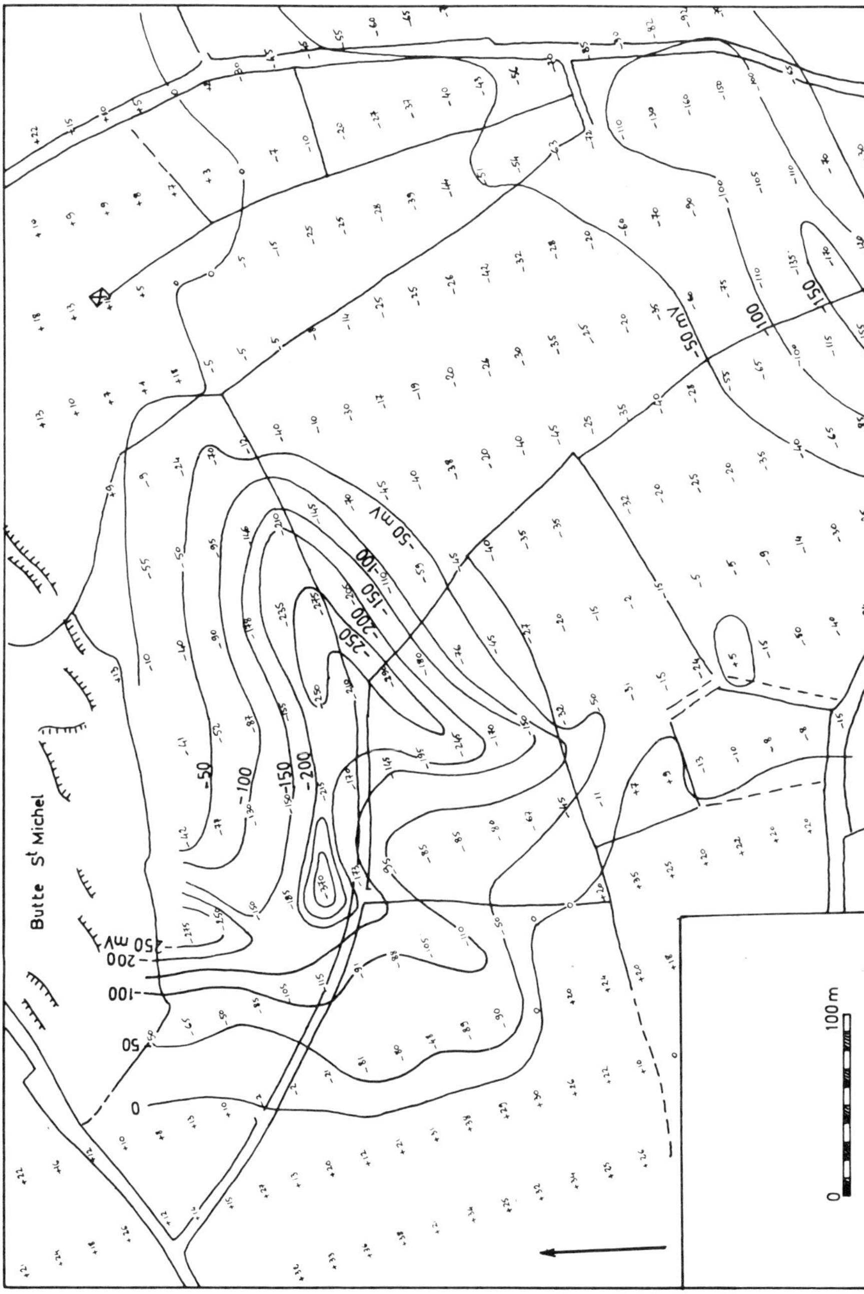

Fig. 5.6. *La Porte aux Moines. SP survey.*

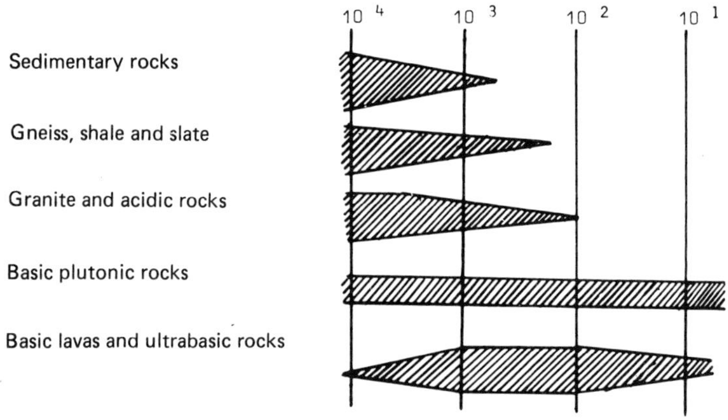

Fig. 5.7. *Magnetic susceptibility of major rock series (in uem CGS).*

These remnant or induced magnetizations cause distortions in the force lines of the earth's magnetic field. The measurement and charting of certain parameters of the magnetic field above the ground surface can thus yield data on the location, geometry and magnetization of deposits underground. These measurements are taken with very sensitive instruments called magnetometers.

It is well known that the earth's field exhibits variable characteristics **(Fig. 5.8)** at different points of the globe. It is directed vertically downward at the magnetic North Pole and directly upward at the South Pole. It is horizontal at the geomagnetic Equator. Its intensity varies from 20,000 to about 70,000nT (nT, Nanotesla, formerly called γ). Its inclination in France is about 60° and its intensity about 45,000nT.

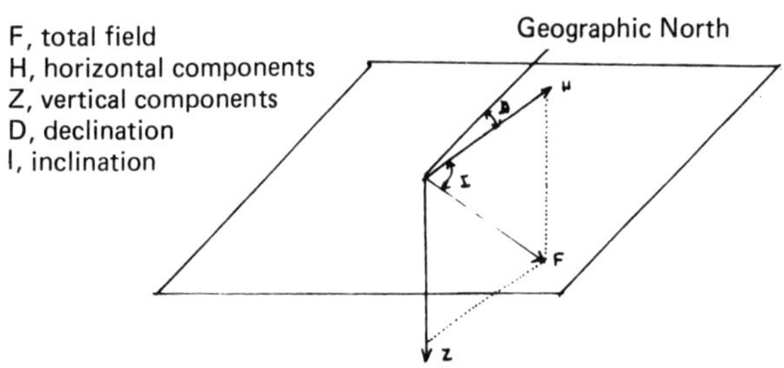

Fig. 5.8. *Characteristics of the earth's magnetic field.*

The knowledge of the approximate characteristics of the local magnetic field is indispensable for the interpretation of the anomalies, whose form depends directly on the inclination of the field.

3.2 Apparatus

Magnetic measurements can be taken on land, at sea, underwater, in drill holes and in the air (by aeroplane, helicopter or even balloon). We only describe two modern types of ground magnetometers here, and ignore the old magnetic balances (Schmidt, Ruska) and even the ABEM torsion magnetometer described in older manuals. These instruments have become obsolete because taking measurements is time-consuming, and the push-button magnetometers of today guarantee a three- to four-fold increase in output.

3.2.1 PROTON MAGNETOMETER

The proton magnetometer measures the total field in line with the following principle. A current is sent for one second in a solenoid surrounding a substance rich in hydrogen nuclei (kerosene, water). The current polarizes the protons (hydrogen nuclei) along the axis of the coil. When the current is stopped, the proton spin is directed along the local magnetic field, assuming a precession motion with a frequency proportional to this field. The emf induced in the coil by this motion is transmitted to the instrument which measures its frequency and converts it into a reading of field intensity, directly displayed on a digital readout to the nearest nanotesla.

Many magnetometers built on this principle are now available, including:

1) G 816 Geometrics, Sunnyvale, (California)
2) MP 2 Scintrex, Toronto (Canada)
3) GM 122 A Barringer, Rewdale, Ontario (Canada)

All these instruments include:

1) the measuring instrument attached to in a waist harness;
2) the coil mounted on a 2.5 to 3m mast and connected to the instrument by cable.

Measurement is extremely simple:

1) with the coil immobilized and the needle pointing approximately North (or South) a knob is pressed and the measurement result is read after a few seconds;
2) two readings are normally taken at the same place, and the mean of the two readings is taken if the readings do not differ by more than 2nT.

3.2.1.1 Precautions

1) Make sure that the batteries are charged. These instruments have a visual testing device which indicates battery strength. When the batteries are discharged the scatter increases abnormally on a given station.
2) Do not carry magnetic objects on your body (coins, knife, compass) and clear the ground of such objects (hammer, machete, 'topofil').
3) Always assume the same position facing North to ensure that the instrument always has the same effect on the coil.
4) Keep the coil stationary and wait until the liquid is stabilized before taking the measurement.
5) Take care when handling the instrument-coil connecting cable or the connections.

6) Note the time of measurement and correct for diurnal variation.

3.2.2 'FLUXGATE' MAGNETOMETER

This magnetometer measures the field component along one direction (usually vertical). The measurement device features two small solenoids with a Mumetal core, positioned side by side. A third solenoid surrounds the first two, conveying an a.c. current whose strength is adjusted so that the cores are nearly saturated. In the absence of an external magnetic field the third solenoid does not convey any current because the two small solenoids are wound in the reverse direction.

The presence of the terrestrial field destroys the symmetry of this device with respect to the hysteresis curve of the solenoid cores, and the third coil then receives a flux proportional to the field component along the axis of the solenoids.

Accuracy may be as good as 1 to 2nT if the instrument is kept very stable vertically. In actual fact, most of these magnetometers are designed to be used quickly, hence without the need for a tripod. Accuracy is usually 10 to 20nT. Many 'fluxgate' field magnetometers are available on the market including:

1) MF1 and MF2 built by Scintrex, Toronto (Canada)
2) MFD4 built by Scontrex, Toronto (Canada)
3) MV-1 built by Phoenix, Toronto (Canada)

Measurement is as simple as with the proton magnetometers. In this case it suffices to keep the instrument vertical, depress a knob and to read the measurement on a pointer scale (MV-1) or digital readout (MFD4). Of the four instruments mentioned above the MFD4 digital magnetometer is the simplest because it requires no zone adjustment or choice of measurement range.

The precautions to be observed are the same as for the proton magnetometers, together with the strict immobility of the instrument in the vertical position during the measurement.

3.3 Measurement system

3.3.1 STATION NETWORK

As for SP the layout of measuring points or stations is very important, because the grid must be sufficiently tight to detect the anomalies explored, but not so tight as to raise the cost of the survey.

All the operator's skill lies in the proper choice of the space between stations, and also in the decision to install extra stations or extra sections for a better definition of an anomaly. Various alternatives are available:

1) Map to be surveyed: stations must be set up on characteristic boundaries (diagonals of plots, roads, etc.) or on traverses already staked out by other methods (geochemistry).
2) Sections: these may be obtained across a structure, a geochemical anomaly, an anomaly detected by airborne instruments (mag or EM), etc. Make sections more or less perpendicular to the structure, while avoiding an

orientation too close to EW (with an equivalent number of stations, the cartography obtained is inferior to that achieved with submeridian sections).

3.3.2 CORRECTION FOR DIURNAL VARIATION

The magnetic field is not constant at any given point. It undergoes periodic and irregular variations which reduce the effective accuracy of the measurements taken and must be corrected wherever possible **(Fig. 5.9)**:

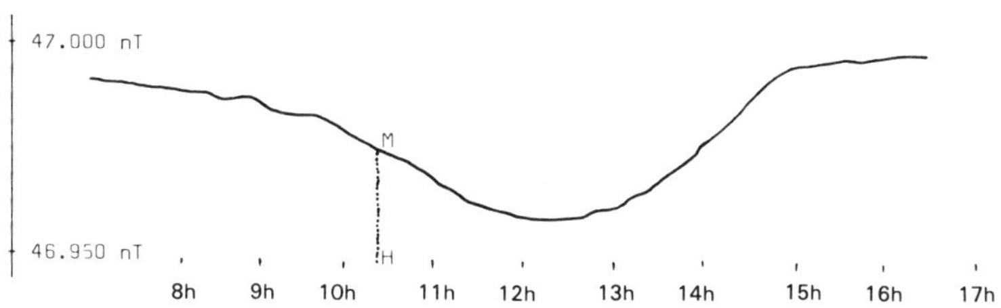

Fig. 5.9. *Example of diurnal variation of the total magnetic field (France).*

1) Short variations (a few seconds) are unimportant (1 or 2nT) and can be eliminated by repeating the measurements at each station;
2) Variations 'within a period of a few minutes' have an amplitude of a few nanotesla but are more difficult to correct because one cannot remain for several minutes at each station without seriously jeopardizing output, and it is difficult to determine accurately on a recording the moment at which these variations occur.
3) The diurnal variation is due to solar activity. It is manifested by a positive or negative deflection of 20 to 40nT between 9 a.m. and 4 p.m. (maximum deflection around noon).

If the prospector is fortunate enough to have one recording station at less than 200 or 300km from the sector to be prospected, it suffices to record the hour and minute of each measurement and to wait to have the recordings to make the corrections. If not, it is necessary to set up one or more 'bases' on the sector surveyed, and to pass through one of these bases each hour. It can be assumed (as an approximation) that the field has varied uniformly between two successive passages through a base. This amounts to plotting the diurnal variation curve with one measurement per hour. Obviously, less frequent passages at the base jeopardize the accuracy of this correction, whereas more frequent passages reduce output.

The base (or bases) must be selected in advance in order to plan the paths on the sections or traverses. A base must be identified very clearly (so as to take measurements at the same place each time). It must be readily accessible, placed at the intersection of several traverses and, if possible, not on an anomaly (unfortunately, this cannot be respected immediately).

The correction for a measurement is made as follows:

1) The diurnal variation curve at the base is first plotted for the day of measurement (Fig. 5.9).
2) A reference line is selected, such as 46,950nT (this value must remain the same for the entire survey).
3) The quantity MH is determined for the time of measurement (10.20 a.m.).
4) This quantity is subtracted from the measurement.

Variations in total field and the variation in the vertical component are not the same. Hence the measurements taken for a vertical component must not be corrected with the use of a total field recording.

3.3.3 MAGNETIC STORMS

Magnetic storms occur on certain days. These cause sudden changes in the magnetic field, which can exasperate a conscientious scientist. It is best to stop taking measurements in the field, to leave the magnetometer at a station and to take measurements every five minutes to check that a magnetic storm has actually occurred. It is first necessary to check that the instrument's batteries are in good condition.

3.3.4 SURVEYS WITHOUT CORRECTION OF DIURNAL VARIATION

When large anomalies are explored (eg 100 to 200nT) or when a vertical component magnetometer is used without a tripod (accuracy of measurement is around 20 to 30nT), there is no point in making corrections of diurnal variation because this correction is extremely small in comparison with the anomalies.

Table 5.4. *A measurement sheet.*

Survey: NIVERNAIS — Section 3 — 25 September

Base B2 45 273 7.15 a.m.

Station	Time	Reading	Correction	Observations baseline
0	08.05	45 381	−10 45 371	
20	08.07	45 391	−10 45 381	fence (barbed wire)
40	08.10	45 400	−11 45 389	
60	08.13	45 390	−12 45 387	
...	......	......	....	

It is necessary to do the following:

1) to locate, on a map, the positions of the sections or stations at the time of measurement;
2) to observe changes in readings so as to take intermediate measurements in case of a sudden irregularity;
3) to record carefully the presence of power lines, metallic fences and other metal objects liable to affect the readings.

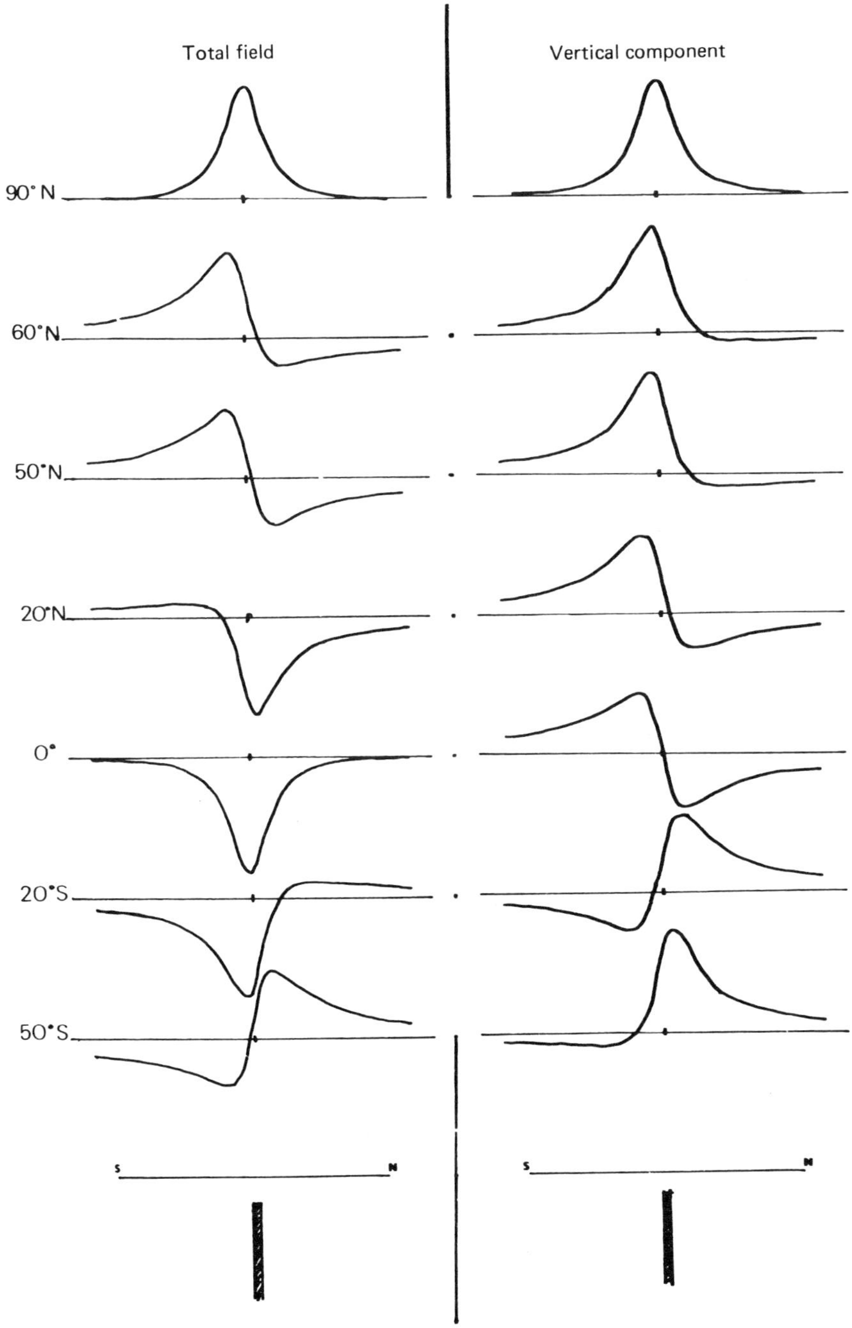

Fig. 5.10. *Anomalies due to a thin vertical vein.*

3.3.5 TRANSFER TO A MAP OR CROSS-SECTIONAL DIAGRAMS

The corrected values can be transferred:

1) to a map if the measurements are uniformly distributed. The values at each station are transferred and isoanomaly contours plotted at a suitable intervals (not too small to avoid too many contours, nor too large so that the anomalies disappear);
2) to cross-sectional diagrams if the measurements have been taken on sections that were wide apart. Use suitable scales to make sure that the anomalies stand out normally **(Figs 5.10 and 5.11)**.

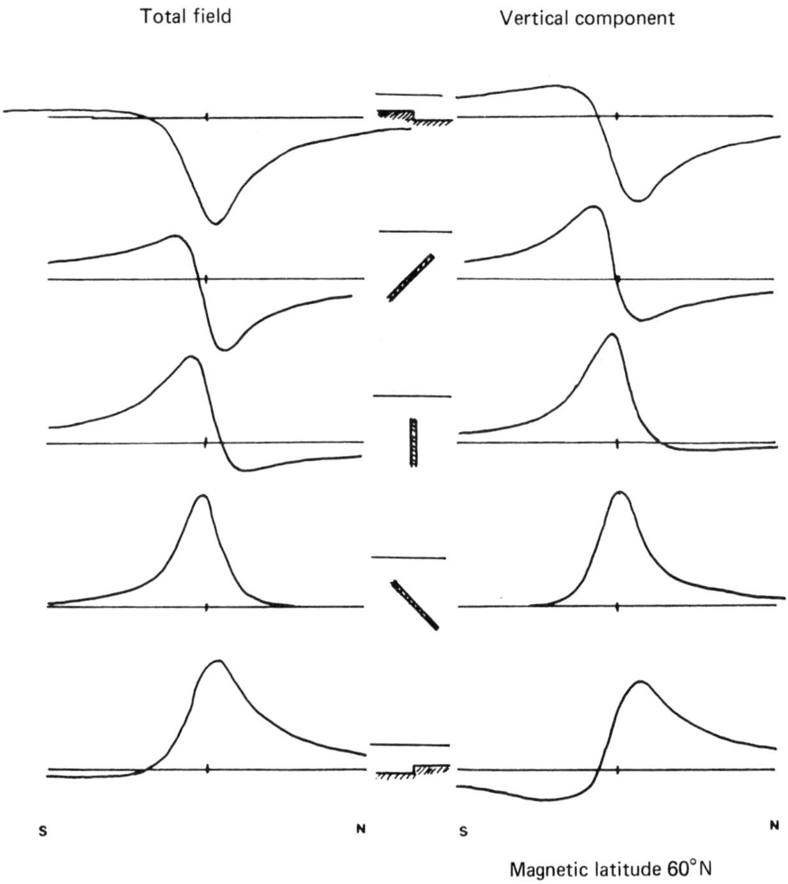

Fig. 5.11. *Influence of dip on the form of the anomaly.*

3.4 Interpretation of maps and sections

The interpretation of magnetic anomalies is a very delicate matter, and should only be performed by experienced geophysicists. It is necessary for the prospector to make quite sure of this, so as to avoid drawing dangerous conclusions for the layout of surface works. In fact, the shape of an anomaly depends on many factors:

1) the form of the deposit, its dip and bearing;
2) the inclination of the magnetization vector (magnetic latitude and any remnant magnetization).

Moreover, the anomaly displays a different shape depending on whether the total field or the vertical component is measured.

One example is the case of an anomaly due to a thin vertical vein, according to the magnetic latitude and the measured parameter **(Fig. 5.12)**. Another example

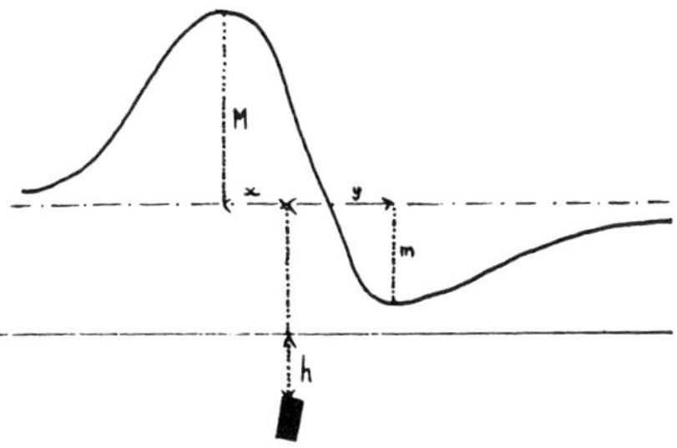

Fig. 5.12. *Bipolar anomaly.*

(Fig. 5.13) shows the influence of dip on the form of the anomaly. Note that the case of zero dip is equivalent to a vertical fault throw.

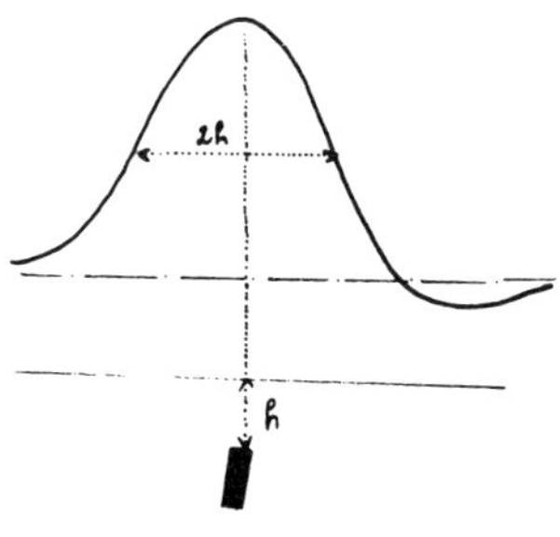

Fig. 5.13. *Unipolar anomaly.*

The prospector should not be thwarted by the pitfall of anomaly bipolarity. It suffices to remember that the magnetic field is not necessarily under the maximum (or minimum) of the anomaly. A simple rule serves to locate the vertical component of the magnetic field under the maximum and the minimum:

This gives approximately $Mx = my$ and $h = \frac{x + y}{2}$

Another trap worth avoiding is the estimation of the depth (d) of the top of the mineralized deposit:

1) if this depth is much greater than the thickness of the deposit the following approximations can be (Fig. 5.13);
2) if the thickness of the deposit is not negligible in comparison with its height, the application of these formulae causes the depth to be overestimated.

An apparent paradox of which the prospector must be forewarned is the weakening of anomalies due to submeridian fields. A single vein listing EW will yield an anomaly which is far more intense than if it was lying NS. In this case, an anomaly would be observed at both the North and South ends of the vein, with practically nothing on the limbs **(Fig. 5.14)**.

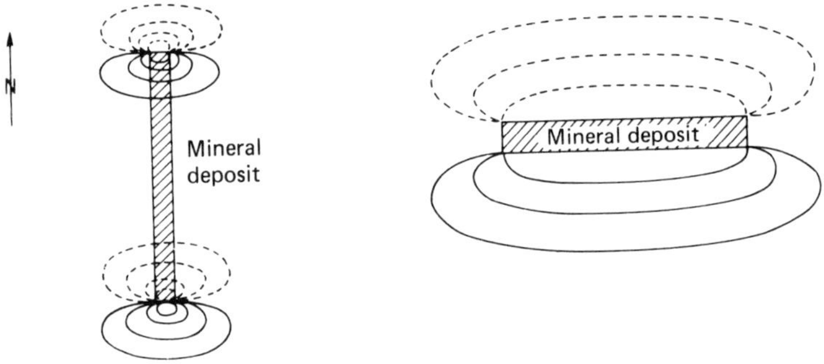

Fig. 5.14. *The magnetic field.*

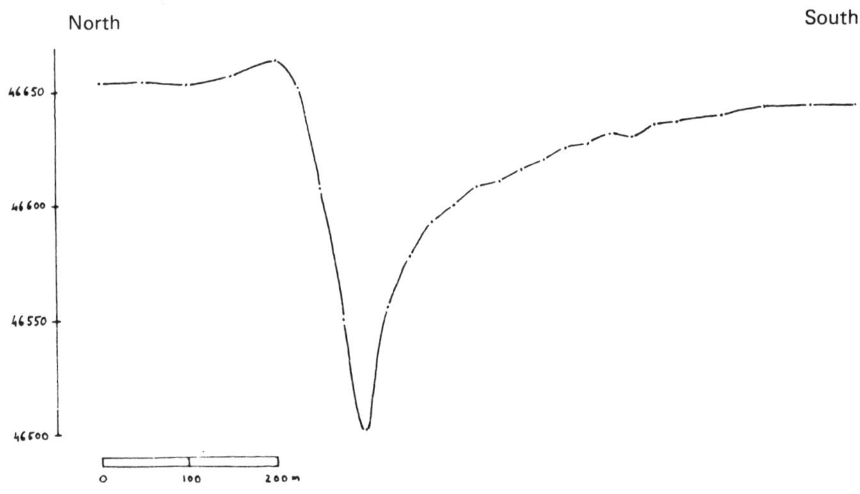

Fig. 5.15. *Total magnetic field in profile.*

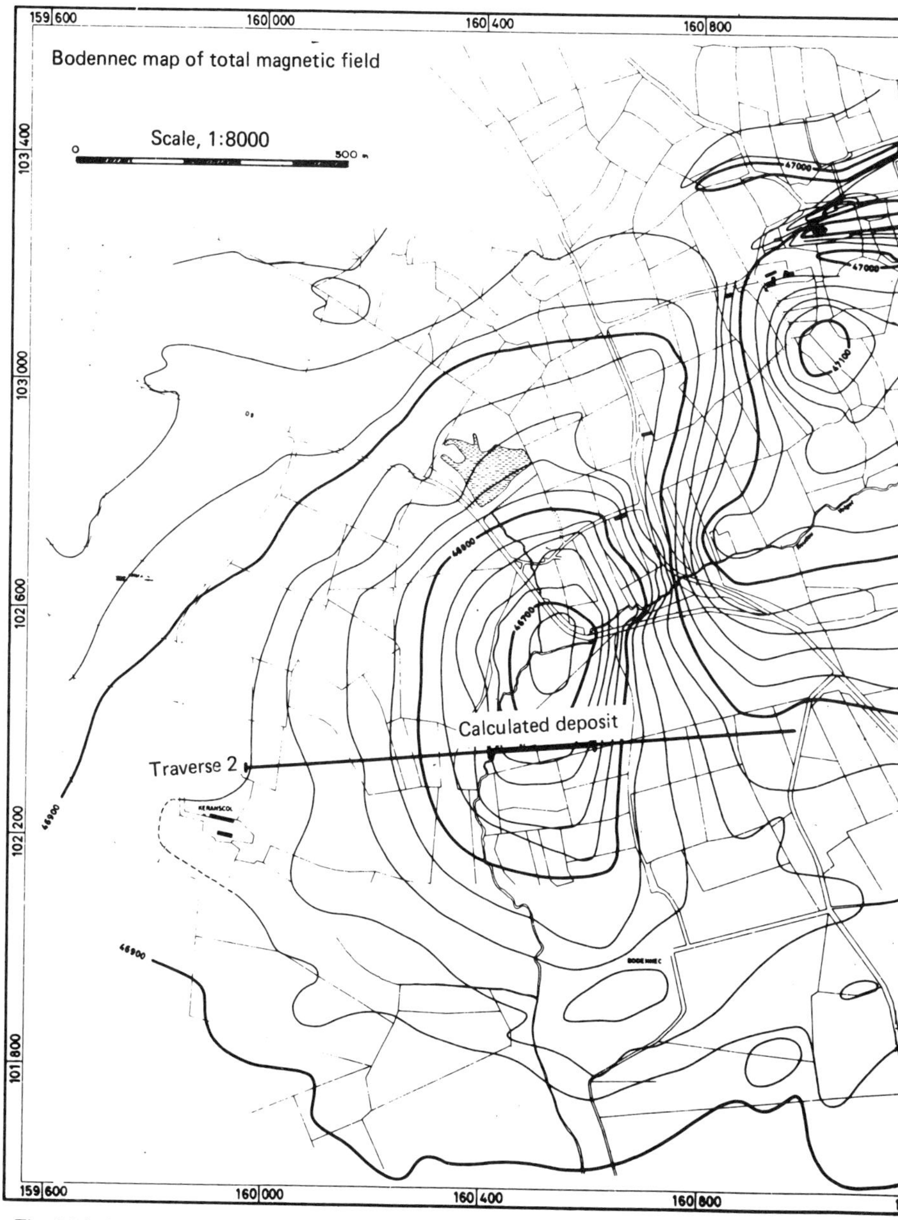

Fig. 5.16. *Map of the total magnetic field.*

3.5 *Examples of magnetic prospecting*

3.5.1 MAGNETIC FIELD IN PROFILE

The form of this NS traverse through France appears to be due to a thin EW vein.

However, it cannot be interpreted by the nomographs in Fig. 5.11; no dip corresponds to the form of the anomaly. On the other hand, Fig. 5.10 shows that the anomaly could be due to a vertical vein with a magnetization direction of 20° upward; hence this implies remnant magnetization. The applications of the depth estimation rules serve to estimate the depth of the top of the vein at 50m (Fig. 5.15).

3.5.2 MAGNETIC MAP OF BODENNEC

This is an aerial test which showed the existence of a magnetic anomaly at Bodennec **(Fig. 5.16)**. The ground survey was designed to pinpoint it very accurately. The result is very complex:

1) the negative pole is SW and the positive pole is NE but they are too widely spaced to correspond to the same source;
2) a submeridian field (fault with throw?) separates these two poles;
3) the negative pole (SW) may correspond to a submeridian magnetic field with a calculated width shown on the drawing found at about 200m below the deposit;
4) the E–W bands in the NE corner of Fig. 5.16 are due to outcropping amphibolite beds.

These two examples illustrate the difficulty of interpreting complex anomalies.

4 Very low frequency (VLF) electromagnetism

4.1 Principle of the method

When an electromagnetic wave penetrates the ground it induces eddy currents in conductors which are out of phase in comparison with the primary field and are distributed in accordance with the geometry of the conductors. These induced currents create a secondary field in quadrature with the primary field. The resultant field (combination of primary and secondary fields) is thus polarized elliptically and is distorted in the vicinity of the conductors. These deformations, measured at the surface, provide information about the presence of underground conductors (eg beds, masses, veins, faults).

Electromagnetic (EM) methods hence require a transmitter (which creates the primary field) and a receiver (used to measure certain parameters of the resultant field). The VLF-EM method uses EM waves emitted by radio stations which are often quite remote (1000 to 2000km away) but very powerful (300 to 1000kW), and are used to send messages to submerged submarines. To penetrate the sea these waves must have a very low radio frequency (15 to 25kHz) and hence a long wavelength (12 to 20km). This means a very low radio frequency, but a high geophysical frequency explaining why the depth of investigation does not generally exceed 40m. On the other hand, this method offers the advantage for the prospector of discarding the transmitter (representing saving in time and personnel) and above all of the emitter at a very great distance and hence operation with 'plane' waves (facilitating interpretation because of the absence of a source effect **(Fig. 5.17)**.

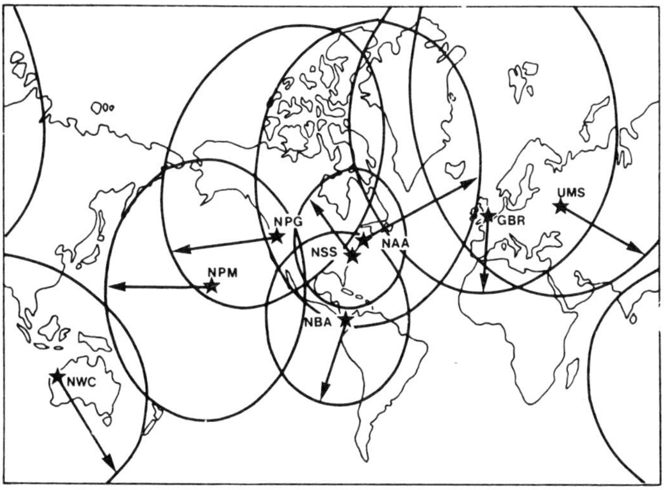

Fig. 5.17. *Major VLF transmitters in the world.*

4.2 Apparatus

Several VLF field units are available on the market, all of them Canadian and manufactured in Toronto (Ontario):

1) EM 16 built by Geonics, Mississauga
2) RADEM built by Crone, Mississauga
3) SCOPAS built by Scintrex, Concord
4) VLF 2 built by Phoenix, Willowdale

In our opinion, the most practical and best-designed instrument is the EM 16 by Geonics. It can also be coupled with a small ancillary unit (Radiohm) which serves to measure apparent resistivity and phase difference. This is the instrument described here.

4.2.1 PROPERTIES OF EM WAVE PROPAGATION

The prospector must be conversant with these properties in order to understand the operations performed. An EM wave propagated in a vacuum possesses two components (magnetic field and electric field), both perpendicular to each other and to the propagation direction **(Fig. 5.18)**.

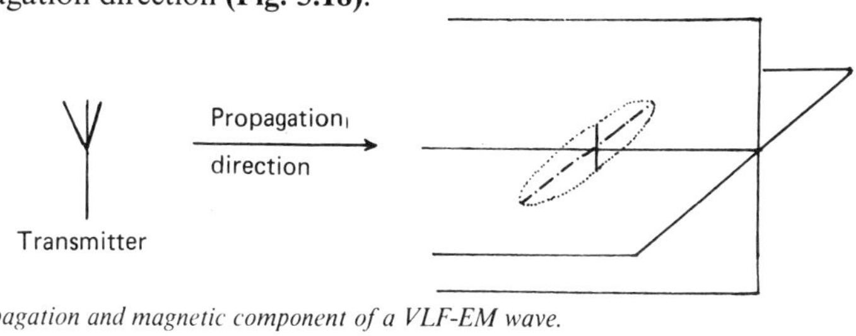

Fig. 5.18. *Propagation and magnetic component of a VLF-EM wave.*

VLF waves are channelled in space between the ionosphere and the surface of the ground. Hence the conditions are not the same as for an EM waves in a vacuum. The magnetic field can be symbolized by a vector whose extremity describes a very elongated ellipse. This ellipse is located in a vertical plane perpendicular to the propagation direction, and its major axis is parallel to the surface of uniform ground **(Fig. 5.19)**.

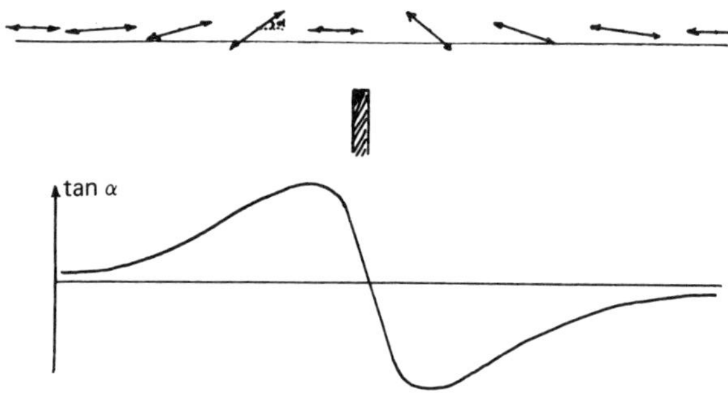

Fig. 5.19. *Variation of inclination of the field near a conductor (the transmitter is in a direction perpendicular to the plane of the figure).*

Thus the magnetic component tends to 'match' the contours of conductors. This clearly explains the occurrence of tilting of the field above a conductor. It is in this way that conductors are detected.

A more detailed explanation utilizes the composition of the primary field and the secondary field created by currents induced in the conductor by the primary field. This reveals the predominant role played by the streams of current flowing through the upper part of the conductor.

Another concept also needs to be clarified, that of the attenuation of the wave of frequency F when it penetrates ground with resistivity (ρ).The wave amplitude decreases exponentially as a function of depth and an attenuation factor is defined as $p = 500\rho/F$ (p in m, ρ in ohm.m, F in Hz) which is the depth at which the wave amplitude is divided by e=2.718. This depth is often inappropriately called the 'penetration depth' (whereas penetration is infinite) or the 'investigation depth' (whereas this is merely an attenuation parameter). However, the prospector can use this formula, remembering that the investigation depth varies with this parameter, but must not be treated as the same thing.

4.2.2 OPERATION AND USE OF THE EM 16 INSTRUMENT

4.2.2.1 Parameters measured

1) Inclination of the major axis of the polarization ellipse;
2) ratio of axes of this ellipse.

These two parameters are equivalent to the phase and quadrature components of the vertical alternating magnetic field (expressed as a percentage of the horizontal primary field), respectively but only if the anomalies are weak.

4.2.2.2 Preliminary operations
The following must be carried out before any other operations:

1) insert the power supply batteries: six 1.5V pen-light batteries;
2) plug in the modules corresponding to the transmitters selected.

This can be done by unscrewing the cover of the instrument to obtain access to the battery cases and connections to plug in the modules.

4.2.2.3 Choice of transmitting stations
FUO 15.1kHz Bordeaux
GBR 16kHz Rugby (GB) for France
NAA 17.8kHz Cutler (USA)

If an approximate idea is available of the elongation of the conductors explored, a station is selected in this direction and sections are obtained perpendicular to this direction.

If no information is available, two surveys are necessary with two stations situated in two more or less perpendicular directions. On the instrument, one connection is available to plug in two modules and a selector to operate with one of the two stations selected.

4.2.2.4 Startup
Flip the switch. A sharp sound should be heard. If not, check for one of the following:

1) the 'volume' knob is on 0: increase the volume;
2) the transmitter is inoperative: turn the selector to find another operational transmitter;
3) the batteries are exhausted: press the small red push button. A sharp sound indicates that the batteries do not require replacing.

4.2.2.5 Performance of a measurement
Having selected a transmitting station and a suitable direction for the traverse the operator is ready to take measurements. He must assume a position on the traverse and take all measurements in the same direction and at regular intervals.

The operator tilts the instrument in the plane perpendicular to the wave propagation direction (which may be slightly different from the plane passing through the section). When the handle of the instrument (which is a ferrite core coil) is perpendicular to the major axis of the polarization ellipse, the instrument is in the minimum reception position because only the component along the minor axis of the ellipse is measured.

This minimum reception can be pinpointed by turning the graduated knob, with the following readings:

1) in the eyelet of the clinometer: the slope of the field (note the sign + upward, - downward),
2) on the rotating knob: the ratio of the axes of the ellipse (also note the sign,

which corresponds to the direction of rotation of the alternating magnetic field vector).

4.2.2.6 Remark
The display of x% on this knob corresponds to the sampling of Nx%/n of the signal received in the small solenoid (n turns) perpendicular to, and at the end of, the instrument handle. This portion of the signal is then dephased by 90degrees and opposed to the signal received in the large solenoid (N turns). If reception is minimal, the following positions are obtained:

1) the large solenoid is parallel to the minor axis of the ellipse;
2) the small solenoid is parallel to the major axis of the ellipse;
3) both solenoids are perpendicular to the direction of the transmitting station;
4) the rotary knob gives the ratio of the axes of the ellipse directly as a percentage.

4.2.3 OPERATION AND USE OF THE EM 16 R INSTRUMENT

4.2.3.1 Parameters measured

1) Apparent resistivity ρ a = K $(Ex/Hy)^2$.
2) Phase shift between Ex and Hy.

Ex is the radial component of the electric field (ie in the station direction). Hy is the tangential and horizontal component of the magnetic field.

4.2.3.2 Preliminary operations
Similar to the EM 16, but with the addition of the an extra instrument (Radiohm), the instruments are connected together with the cables of the two electrodes, and selector 16 is set to 16R. Set the station selectors (EM 16 and Radiohm) to the appropriate position.

4.2.3.3 Choice of stations
Unlike the EM 16 it is necessary to select a station in a direction perpendicular to the elongation of the conductors explored. The sections are therefore in the direction of the transmitter selected.

4.2.3.4 Use
The electrodes are inserted on the section 5m on either side of the operator and aligned in the direction of the transmitter: check the direction. The instrument is held so that the small solenoid is horizontal and perpendicular to the direction of the transmitter.

The minimum sound must then be obtained by turning the knob of the EM 16 and the knob of the Radiohm in succession. The following readings are then taken:

1) on the red graduation of the EM 16 knob, an indication which, multiplied

by the factor (10, 100 or 1000) selected by the scale switch, gives the apparent resistivity in ohm.m;
2) on the Radiohm knob: the phase in degrees.

On uniform ground, the true resistivity is obtained, with a phase shift of 45°.

4.3 Measurement system: station and section network

4.3.1 MEASUREMENT SYSTEM

It is important to remember that to detect a structure with a presumed bearing X, it is necessary to use sections which are approximately perpendicular to X **(Fig. 5.20)**:

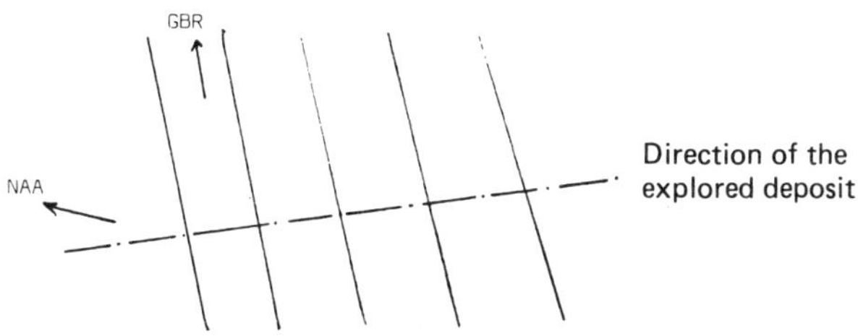

Fig. 5.20. *Layout of a network of sections to be surveyed with the EM 16 (with NAA) and EM 16 R (with GBR).*

1) with EM 16: with one station approximately in the X direction;
2) with EM 16 R: with one station situated approximately perpendicular to direction X.

As a rule, both EM 16 and EM 16 R can be used. If both methods are used, it is wise to plug two modules into the EM 16 instrument, corresponding to two stations located in approximately perpendicular directions. The term 'approximately' is intended to mean within 30 to 40°.

Using the EM 16 it is necessary to proceed on parallel sections with stations regularly spaced (10 or 20m).

Using the EM 16 R operations can be performed on sections as with the EM 16 or with stations distributed more or less uniformly on the sector explored. The apparent resistivity values can then be transferred to the section or to a map, and in the latter case, the isoresistivity curves can be plotted to reveal the position and form of the resistivity or conductors.

However, transfer of EM 16 measurements to a map or section does not allow easy interpretation, and a rough processing of the measurements can make up for this drawback.

4.3.2 TREATMENT OF EM 16 MEASUREMENTS BY DERIVATION

Figure 5.21 shows the shape of an anomaly of the inclination of the field obtained with the EM 16 on a vertical conductor. The ellipticity anomaly should be identical

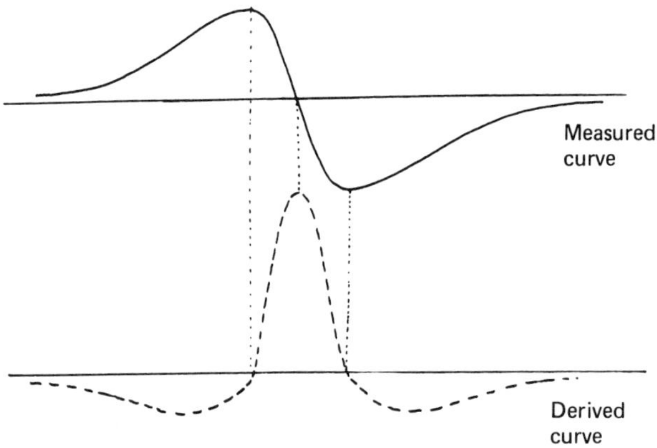

Fig. 5.21. *Derivation of the measured curve.*

in shape. This structure is located directly below the inflexion point, which is never very easy to identify on a less regular section or on a isoanomaly contour map.

This is why the curve is subjected to the following:

1) a derivation which serves to convert the inflexion point to a more easily perceptible peak;
2) smoothing to eliminate the irregularities in the curve.

For derivation it is sufficient to take the difference between two consecutive measurements a and b and divide this by the interval $\overline{AB}$, and to multiply the value found at the centre of $\overline{AB}$. If care has been taken to ensure the same interval between the stations and to take measurements in the same direction, it is unnecessary to divide by $\overline{AB}$.

Fraser describes a simple method to perform derivation and smoothing simultaneously. Starting with four consecutive measurements a, b, c and d the quantity (a+b) – (c+d) is calculated and assigned to the middle of segment BC. To prepare for this operation the field log is present slightly differently.

4.3.3 FIELD LOG

As for SP and magnetism, we recommend recording the measurements on a loose-leaf pad, with the sheets safely stored once they are filled. When using the EM 16 R it is only necessary to fill in columns ρ and π without any subsequent treatment.

Do not forget to record the positions of the sections and the stations carefully on a map. Correct identification avoids the need to return to the field and contributes considerably to the quality of the results.

4.4 Interpretation of results

The interpretation of VLF results is generally qualitative because, given the slight

Table 5.5. *A field log sheet 'VLFEM 16'.*

Survey: LA CHATAIGNERAIE					18 May
		Section 3			
Transmitting station	tan (INF) NAA	(π) GBR	(ρ) GBR	(ε) NAA	Observations
0	0		−1		Baseline
10	+1		0		
		−12		−6	
20	+3		+1		
		−7		−3	
40	+1		0		
		+24		+9	Telephone line
60	−6		−2		Hedge
		−10		−4	
70	−3		−1		
80		0		0	

investigation depth of this method (about 40m), it is rarely necessary to estimate the depth of the cause of an anomaly. This can nevertheless be achieved with the results of EM 16.

4.4.1 EM 16

The form of an anomaly obviously depends on the shape of the conductors. **Figure 5.22** offers examples of anomalies on vertical conductive veins (thin or thick). The case of contact between a conductive compartment and a resistive compartment is easily inferred from the case of the thick vein. Any dip introduces an asymmetry in the section curves.

It is important to stress the case of anomalies created by the summit of a ridge or the bed of a gully. These anomalies are dangerous pitfalls, especially since ridges and thalwegs often have geological significance.

Information may be available on the resistivity contrast by the comparison of the inclination and ellipticity curves:

1) high resistivity contrast: the inclination and ellipticity vary in the opposite manner;
2) low resistivity contrast: the inclination and ellipticity vary in the same way (Section 4.5.3).

In the case of a thin vein or a horizontal linear conductor (conducting pipe or cable) the depth may be estimated as half of the interval between the maximum and minimum of the rough inclination curve of the field.

4.4.2 EM 16 R

If the compilation of the 'apparent resistivity' parameter yields data on the geometry of the structures (conductive or resistive), the 'phase' parameter may, in the case of horizontally bedded formations, offer information on the thickness of the first formation encountered:

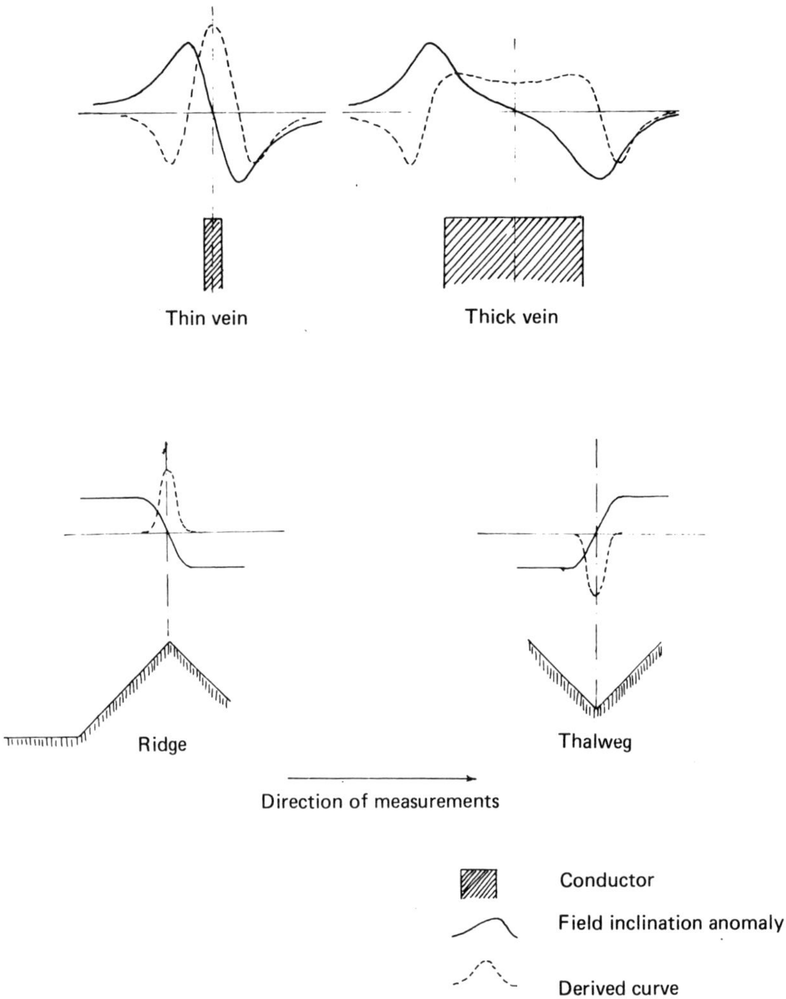

Fig. 5.22. *Some types of VLF anomaly.*

1) First case: $(\pi) = 45°$, the first formation has a thickness greater than the attentuation parameter p (defined in Section 4.2.1).

   $\rho = \pi$ measured.

2) Second case: ø 45°;e<ρ. If ø>45°,$\rho_1 > \rho_2$; if ø<45°,$\rho_1 < \rho_2$; ρ_1 and ρ_2 are the resistivities of the first and second sections.

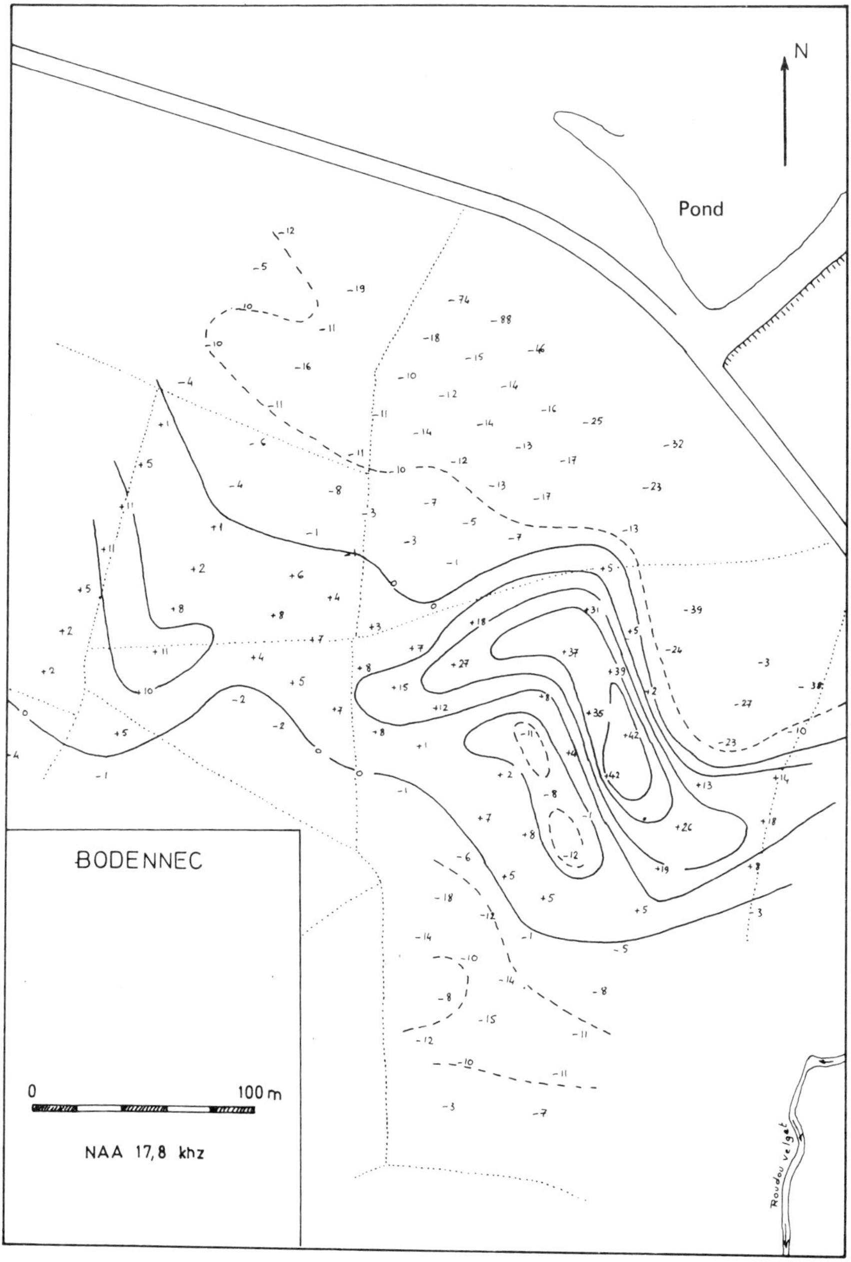

Fig. 5.23. *VLF-EM, derived and smoothed values of field inclination.*

4.5 Examples of VLF anomalies

4.5.1. ANOMALY DUE TO A LENS OF MASSIVE SULPHIDES: WEST OF BODENNEC (**FIG. 5.23**, NAA WEST TRANSMITTER)

Isoanomaly contour map of the inclination of the field, derived and smoothed is shown in Fig. 5.23. Note the Z form of the anomaly. This could only be identified by narrowing the sections to 20m. The bore holes drilled in the anomaly encountered (at a depth of about 10m) the top of a lens of massive sulphides plunging northward. The twisted appearance of the anomaly was explained subsequently when it was observed that a second tectonic direction had affected the deposit.

In the SW is a small negative anomaly which normally accompanies the main positive anomaly. At the NE appear high negative values which accompany a very intense positive anomaly running along the road and due to a telephone line.

4.5.2 ANOMALY DUE TO A RESISTIVE VEIN

This is the apparent resistivity parameter measured with EM 16 R, which serves here to identify a resistive N–S vein near Ruffiac (**Fig. 5.24**). The transmitter used

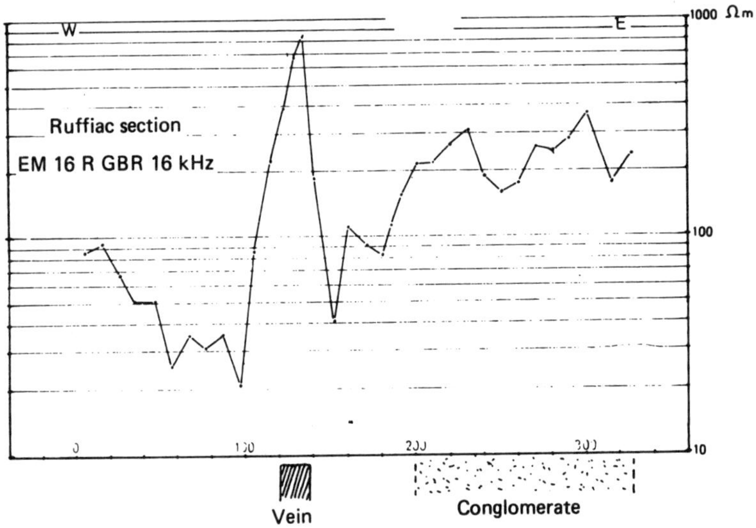

Fig. 5.24. *VLF-EM anomaly due to a vein.*

is GBR located in the North. Note not only the resistivity peak due to the vein, but also the resistive range due to a bed of conglomerates.

4.5.3 ANOMALY DUE TO A FAULT IN BAJOCIAN LIMESTONE (REMOVILLE, VOSGES)

This fault behaves as a medium-conductive vertical plane (probably due to the shale fill) in a resistive environment. The position of the fault is located to the nearest 10m (230 to 240m). The accuracy would have been better with measurements taken at 10m intervals. Note that this location is easier to identify on the derived and smoothed curve. **Fig. 5.25).**

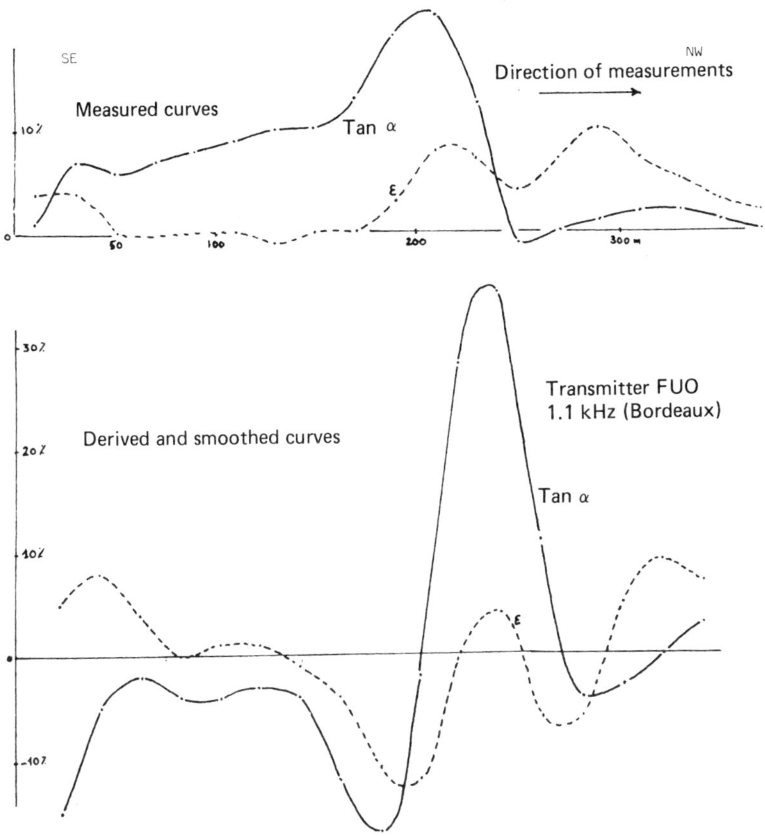

Fig. 5.25. *VLF-EM anomaly due to a fault.*

The fact that ε varies in the same manner proves that the resistivity contrast (between the fault and the surrounding formation) is average **(Table 5.6)**. For comparison we give the value of the resistivity of three resistive minerals:

1) sphalerite: 10^5 ohm.m;
2) stibnite: 5×10^6 ohm.m
3) calcite: 5.5×10^{12} ohm.m

Resistivities of geological formations are given in **Table 5.7**.

Table 5.6. *Order of magnitude of resistivity in ohm.m for different minerals.*

OXIDES		
Magnetite	$Fe_3 O_4$	$5,2.10^{-5}$ à $7,1.10^{-2}$
Hematite	$Fe_2 O_3$	$4,0.10^{-3}$ à $1,8.10^{-2}$
Pyrolusite	$Mn O_2$	$1,8.10^{-2}$ à 50,0
MISCELLANEOUS		
Graphite	C	$3,8.10^{-7}$ à $9,8.10^{-3}$
Native copper	Cu	$1,2.10^{-8}$ à $3,0.10^{-7}$
SULPHIDES		
Galena	Pb S	$1,32.10^{-5}$ à $1,87.10^{-2}$
Pyrite	$Fe S_2$	$1,20.10^{-3}$ à $3,80.10^{-1}$
Pyrrhotite	$Fe_{(1-x)}S$	$2,00.10^{-5}$ à $1,23.10^{-4}$
Chalcopyrite	$Fe_2S_3Cu_2S$	$3,70.10^{-4}$ à $1,40.10^{-3}$
Marcasite	$Fe S_2$	$1,00.10^{-6}$ à $1,50.10^{-1}$
Pentlandite	$(FeNi)_9S_8$	$1,00.10^{-6}$ à $2,10.10^{-5}$
Cobaltite	CoAs S	$2,66.10^{-2}$ à $1,20.10^{-1}$
Mispickel	FeAs S	$3,00.10^{-4}$ à $1,93.10^{-2}$
Molybdenite	MoS_2	$2,00.10^{-2}$ à 5,00
Covellite	Cu S	$1,00.10^{-6}$ à $5,00.10^{-4}$
Chalcocite	Cu_2S	$8,00.10^{-5}$ à $1,00.10^{-2}$
Bornite	$Fe_2S_3nCu_2S$	$2,00.10^{-6}$ à $1,00.10^{-1}$

Table 5.7. *Order of magnitude of resistivities of geological formations.*

Resistivities of electrolytic solutions

Sea water	0/2-0/3ohm.m
Spring water	
(Sedimentary rocks)	10 to 100ohm.m
(Crystalline rocks)	100 to 1000ohm.m
Rain water	1000 to 10,000 ohm.m

The resistivity of rocks is essentially determined by the presence of water, clays and electron conductors and their distibution (porosity, fracturing, etc.).

- Clays: 0.5 to 100ohm.m
- Sands: 1 to 1000ohm.m (depending on the resistivity of the impregnation water and the clay content) > 100,000ohm.m for dry dune sands
- Massive limestones: 2000 to 10,000ohm.m
- Graphitic schists of the Pyrenees: 0.1 to 200ohm.m (depending on graphite content)
- Metamorphic schists: 300ohm.m
- Sandstones: 500 to 4000ohm.m
- Quartzites: 1000 to 10,000ohm.m
- Salt: 10^6ohm.m
- Marble: 10^{12}ohm.m
- Granite: 1000 to 15,000ohm.m (unless weathered)
- Basalts, andesites, peridotites: 800 to 15m000ohm.m (unless weathered)

5 Scintillometry and spectrometry

5.1 *Principle of the method*

5.1.1 GENERAL DESCRIPTION OF RADIOACTIVITY

The decay of radioactive elements produces three types of radiation:

- α emission of helium nuclei;
- β emission of electrons;
- γ emission of 'hard' electromagnetic radiation, ie with a very short wavelength.

α and β radiation are absorbed by a few cm or dm of air and are of no interest to us here. γ radiation, however, which is sufficiently penetrating, can be detected to an altitude of 100 to 200m, and its absorption increases as the relative humidity of the air increases.

Radioactive substances are divided into several groups as shown in **Table 5.8**.

Other radioactive isotopes are found in nature, such as ^{40}K – ^{87}Rb. Each decay occurs with a specific energy (hence over a specific wavelength). Instruments capable of discriminating the different radiations according to their energy are called spectrometers. Those which measure total γ radiation are called scintillometers if the detector is a scintillation crystal or Geiger counters if the detector is a tube.

Table 5.8. *Groups of radioactive substances.*

Group	Source element	γ	emitting isotopes
Uranium	radium	^{238}U	^{226}Ra – ^{214}Bi – ^{214}Pb
Thorium		^{232}Th	^{228}Ac – ^{228}Th
Actinium		^{235}U	

5.1.2 RADIOACTIVITY GENERATED BY ROCKS AND MINERALS

Since rocks display varying degrees of radioactivity depending on their radioactive element content, it is clear that the measurement of γ radiation above outcrops can serve to obtain data on the geology or on the presence of radioactive ores.

5.1.2.1 Radioactive ores

PRIMARY MINERALS

Uraninite, UO_2, in cubic or octahedral crystals, a classic mineral of granitic pegmatites, containing rare earth elements in the associated impurities: Ce, La, Y, Fr and Th up to 14%.

Pitchblende, UO_2, uraninite of a variety in which concretions have been formed, massive and blackish cryptocrystalline, containing very little thorium, easily weathered, in veins or impregnations in granitic massifs. It is also found in microscopic inclusions in sedimentary rocks (bitumens, sandstones, conglomerates).

Coffinite, uranium silicate, black and powdery.

OXIDATION MINERALS

Pitchblende gives rise to a wide range of weathered products: hydrates, phosphates, vanadates, silicates, sulphates, arsenates, etc. Their dominant colour is orange-yellow to red or greenish yellow. Among these many species, we only mention those which yield ores of economic value.

Gummites, $UO_3.nH_2O$, cryptocrystalline minerals pale to orange-yellow in colour, exact composition uncertain.

Hydrate group, orange-red curite (Katanga); billietite, a hydrated oxide of Ba and U, a surface ore of Margnac (Haute-Vienne); fourmarierite, crimson red, a hydrated oxide of Pb and U (Katanga).

Uranophosphate group, Autunite, in pegmatites and granites, pale green fluorescent to ultraviolet, a secondary economic ore; phosphuranylite; uranocircite; renardite; chalcocite, emerald green, combined with copper, a secondary economic ore.

Uranovanadate group, Carnotite, yellow colour (Australia, Colorado); francevilite (Gabon), tyuyamunite (Siberia).

Silicate group, Uranotile honey yellow, soddytite pale yellow, kasolite yellow ochre.

THORIUM ORES

These ores are of the same type as uranium ores. Thorium is sometimes only present, and sometimes combined with uranium, in the form of the oxide, hydrate, phosphate, silicate, and also often combined with rare earth elements.

Thorite, Th silicate, secondary mineral of granitic rocks or found in veins (USA).

Thorianite, Th.U oxide, the main mineral mined in alluvial deposits (Madagascar).

Monazite, a phosphate of rare earth elements, the main ore mined throughout the world for rare earths, in granites, syenites, pegmatites, veins and sometimes stockworks.

Other rare earth minerals sometimes contain Th and U (pyrochlore, fergusonite, samarskite, euxinite).

Figure 5.26 provides an idea of the natural radioactivity of the different groups of rock in accordance with the radioactive decay scheme involved.

5.1.3 BACKGROUND

Apart from any radioactivity anomaly, the instruments can measure a quantity of pulses which constitute the background and is due to the following:

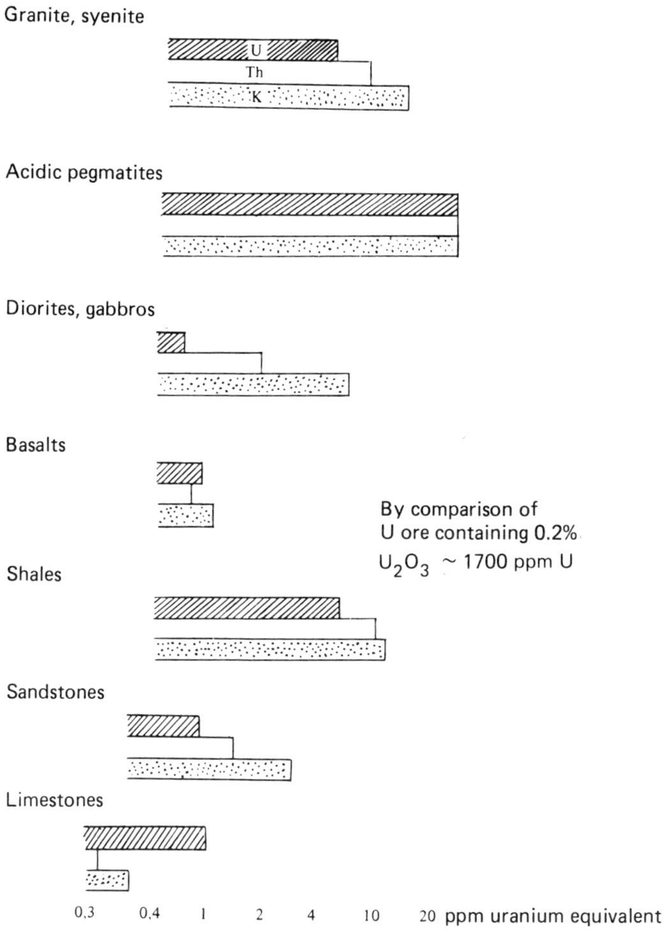

Fig. 5.26. *Comparison of the radioactivity of a few rocks.*

1) cosmic radiation (which depends on the coordinates of the site and the solid sensor interception angle);
2) mean local radioactivity of the rocks and the air.

5.2 Measuring instruments

5.2.1 GAMMA METERS

These instruments operate on the principle of the detection of γ radiation by a Geiger-Muller tube. We only mention them for information because they are now rarely used for ground prospecting. They are slow, sensitive to impact and have the drawback of being sensitive to cosmic rays (high background).

5.2.2 SCINTILLOMETERS

Scintillometers are γ ray detectors. They feature a crystal (stilbene, sodium iodide with thallium, etc.) which scintillates when it absorbs a γ particle. The scintillation is converted into an electrical pulse by a photomultiplier cell. A counting system serves to display the radiation intensity. The measurement is simple and rapid (a few seconds).

5.2.2.1 A scintillometer: SPP 2 NF (Saphymo-Scrat) France, weight 3.5kg
This instrument is easy to handle. It has a single control knob for operation, battery check and sensibility selection. Another knob provides an adjustable acoustic signal.

5.2.3 SPECTROMETERS

These are scintillometers which can discriminate the gamma rays produced by the decay of ^{235}U and ^{238}U, ^{232}Th or ^{40}K, according to their energy.

5.2.3.1 A spectrometer: DISA-300 by Geometrics, Exploranium, USA, by weight: 3.2kg
This instrument is of the integral threshold type, indicating that it can measure radiation above:

1) 0.1Mev: total radioactivity;
2) 1.3Mev: $^{40}K + U + Th$;
3) 1.6Mev: U + Th
4) 2.5Mev: Th

Operation is very simple. Two selectors can be used to choose the count time and the radioactive elements (U, Th or K). Another knob actuates the measurement which is digitally displayed. The instrument is powered by standard batteries.

5.3 *Use of instrument*

In general reconnaissance the geologist has a scintillometer at the base camp and must check all the samples brought in. In certain cases, the prospector carries an instrument equipped with an acoustic signal and makes a ground check if this signal is tripped. In any case, the counter can help to delimit a geological contact between the basement and a sedimentary formation.

5.3.1 SELF-SUPPORTED RECONNAISSANCE PROSPECTING

In France, a country with a high density of secondary and country roads, it is possible to carry out wide ranging, self-supported reconnaissance. The crew must number at least two, one of whom is the guide, who does not deal with the equipment. The prospector has to use an instrument with high sensitivity and rapid reaction to detect an anomaly from a travelling vehicle.

In this case, the vehicle is stopped, the prospector alights to perform a check on foot, and if the anomaly is confirmed, a rapid sampling is performed using the

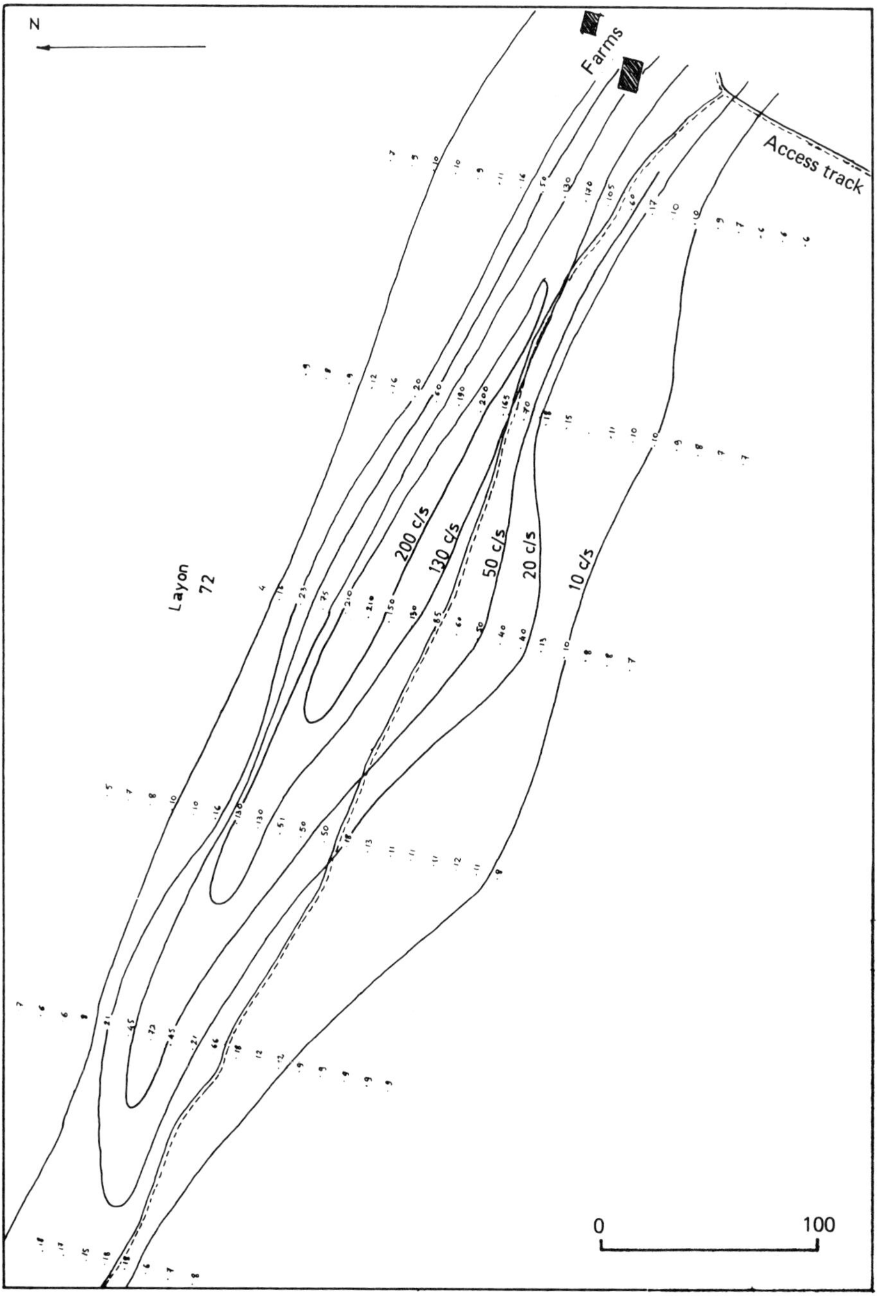

Fig. 5.27. *Sinnamary (French Guyana) count map, scintillometer from Geometrics.*

auger, with measurement by probe. This type of prospecting is also feasible in flat arid zones, in which case large traverses are performed on a 1 to 5km mesh grid. If an anomaly is identified, a count map is drawn up.

5.3.2 COUNT MAP

The count map is a scintillometric (or spectrometric) map of the measurements taken on a given sector with a rather tight grid (5 to 20m). The speed of measurements actually depends on the counting time, which must be adapted to the intensity of the γ radiation observed. No magic formulae can be given for the mesh (square, rectangular, widely spaced sections) or for its dimensions. If measurement time is short, it does not cost much more to take closer measurements (under 20m). If the sector surveyed is vast, the sections can be spaced out to try to identify the anomalies, which are subsequently examined in detail.

Figure 5.27 gives an example of a count map charted on the sands of ancient offshore bars in French Guyana, where the radioactive deposit is monazite. These offshore bars were highly elongated and very narrow, justifying a survey by widely spaced sections (100 to 200m) by measurement points at 10m intervals on these sections.

5.3.3 PRECAUTIONS TO BE OBSERVED

The following must always be indicated on a count map:

1) the type of instrument used and its serial number: the instruments are graduated in c/s and calibrated with varying degrees of accuracy, so that sensitivity depends on the size of the detecting crystal;
2) dates of measurements: this is necessary to put the prospecting in the rainfall context: in the report, the date of the last rains and their abundance must be mentioned, together with the pedological characteristics of the soil (laterite, clay, sand, plant soil). This is because γ radiation is easily absorbed by water and 50cm of wet earth can nearly completely mask an anomaly. Hence the prospector must record the presence of outcrops or, if not, the

Table 5.9. *A page from the field log.*

Survey: beach sands, French Guyana Instrument Geometrics GS 101, No 1298			Prospect: IRACOMBO Date: 16 November
Section B			Observations
0	12	Track	Background
10	12	Plant soil	10c/s
20	15	Plant soil	Heavy rain during the night
30	25	Plant soil	
40	35	Sand	
50	37	Sand	
60	25	Sand	
70	10	Backwater	
80	10	Backwater	

thickness (presumed) and the level of moisture in the soil, if he wishes to be able to interpret his count map.

5.3.4 FIELD LOG AND TRANSFER TO MAP

The field log is easy to record, and should contain the maximum amount of data.

The map must be prepared to a suitable scale and the measurements must be transferred after each day in the field, in order to take extra measurements or establish intermediate sections if necessary.

6 General recommendations

This chapter will enable the prospector and the geologist to use the simples geophysical methods available. However, it is necessary to stress a number of essential recommendations which are valid for all the methods described:

1) Take care of the equipment: the instruments are robust and designed to operate in severe conditions, but this is not a reason to handle them incorrectly such as allowing them to collide with objects, to remain in the rain or sun, or entrust them to unskilled members of the group to perform the measurement themselves. The geologist or prospector who has taken charge of an instrument must be responsible for it and for its storage and upkeep.
2) Geophysical measurements must be positioned:

   - in the field: install topographic markers;
   - on a map: if insufficient, locate the stations in relation to the markers;

   The location of any station must be identified clearly, in order to take extra measurements, to drill bore holes or to check the results of other methods (geochemistry or geophysics).
3) The geophysical results must be interpreted preferably in collaboration with a geophysicist, first because the interpretation is not easy, and also because the geologist or prospector will have a tendency to see in the results only whatever confirms their previous assumptions, and to discard anything that does not agree with what they believe.

References

Geonics Ltd various technical notes. Operation manuals for the EM 16 and EM 16R.

Patterson, N R; Ronka, V (1969) Five years surveying with VLF methods. SEG.

Chapter 6

Mineral exploration drilling

1 Equipment

Drilling is a means of taking samples over a range of depths. The light-weight machines used at shallow depths are often operated by the prospector himself. For greater depths the work is done by a specialist drill crew, the role of the prospector being essentially to supervise the recovery of samples, lay them out, examine them, have analyses made of those which he considers interesting and to make out a drill log.

The devices used are described briefly in order of their use in the course of mineral exploration.

1.1 Light-weight machines

These consist of hand or motorized augers, some of which have already been described in Chapters 3 and 4, for sampling in geochemical alluvial prospecting and prospecting littoral formations, where they are used as much in exploration as in semi-systematic and systematic prospecting.

Some of the light-weight motorized augers worth mentioning are the Bonne Esperance, the Mobile Drill and Minuteman. Medium power motorized augers also axist, such as the Pendrill, Mobile Drill B30, B40 or B50; CME 45 and SIMCO, which require the presence of a driller.

The hand or motorized 'Banka' drill has already been described in Chapter 3. It is used mainly for the exploration and measurement of alluvial or eluvial deposits.

More recent arrivals on the market are the percussion or vibrating drills, used in prospecting for sand and gravel at sea as well as on land, such as the electric V 498 vibroprobe developed and used by the BRGM for prospecting the continental shelf.

All of these machines, whether light or heavy, manual or motorized, are used only in the exploration for and study of friable prospects or deposits (alluvial or eluvial deposits, especially in littoral formations). The prospector may, however, need to use light-weight and portable coring drills himself on stable ground. These will be mentioned later.

1.2 Drills

These are of two kinds: destructive drills and coring drills.

1.2.1 DESTRUCTIVE DRILLS

These are machines with percussion bits that break the rock into fragments (cuttings) which are brought to the surface by circulating air or water, where they are then collected. There are two types of percussion drill:

1.2.1.1 Wagon drills

Wagon drills are conventional mining hammer drills, mounted on a slide on which the hammers move vertically; this is itself normally mounted on a wheeled carriage. The hammers strike a string of drill rods which are screwed together and have a tungsten carbide insert or button bit at their lower end. These mobile rapid advance low cost machines are inconvenient in that they give only relatively few cuttings (about 5kg/m). The working depths achieved vary from 30m (maximum 50 to 60m) for light-weight wagon drills (Atlas BVB or Montabert CFL types) to 70 to 80m (maximum 100m) for the big Atlas Roc 601 type wagon drills. Efficiency for 40m holes varies from 100 to 150m per 8h shift, depending on the ease of access and site changing.

1.2.1.2 Down-the-hole hammer drills

In the down-the-hole hammer system, unlike the wagon drill, the perfectly cylindrical and smooth percussion hammer is at the bottom of the hole. It is rotated by hollow rods which supply the hammer with compressed air that drives it. The average depth reached is 80 to 100m. The diameter of the hole, 65 to 170mm, is larger than that of the wagon drill, which is 40 to 60mm, leading to a greater recovery of cuttings.

Drills of this kind can be classed as light hammer drills (survey, Ingersoll type), capable of reaching depths of 300 to 1000m, that can be attained by casing the bore hole. At present the best casing equipment, giving a contamination free recovery, is made by Stenuick.

The efficiency of holes less than 300m away from one another is around 80m for a 10 to 11h shift, with one driller and two or three assistants, including setting up time. The cost per metre is around three to four times more than with the wagon drill and two to four times less than with the diamond drill.

For the down-the-hole hammer drills, as for wagon drills, dry cuttings recovery takes place in a cuttings collector. The best system consists of a vacuum-operated static cyclone made from an accumulation and recovery cone welded on to the base of a 200 litre drum. It can also be a wet process. When the cuttings are wet and stick to the walls the hammer drills are able to inject or allow water to circulate washing the walls and raising the sludge. If a lot of water is introduced, but not under too much pressure (100 litres/m or less), recovery is possible but the fines will escape and the proportional representation will be distorted. Cuttings recovery from percussion drills presents difficulties that have yet to be overcome so research is in progress to find ways of carrying out *in situ* analyses using geophysics and electronics.

1.2.2 CORING DRILLS

These are machines capable of taking from the ground cylinders of rock called 'cores' which are then carefully examined by the geologist or prospector. The most popular machines are drills with a diamond coring bit, using rotation and pressure to cut out the core.

1.2.2.1 Diamond bits

The bits may comprise either stones of approximately 1mm diameter, set into the surface of an abrasion resistant material, or much finer stones uniformly spaced out in a matrix which wears down revealing more diamonds as the bit is working; the latter type is used particularly in hard formations. The bit is generally topped by a reamer shell with diamond studded inserts or collars which act as stabilizers for the drilling bit and keep the hole drilled at a fairly constant section.

1.2.2.2 Core barrels

The bit sleeve assembly is screwed on to the core barrel, a tube that acts as a core receptacle, and of which there are different kinds.

Single tube core barrels are made up of one tube serving both as an extension of the bit body and as a core receptacle.

In double tube core barrels the inner barrel, mounted on bearings, serves as a non-rotating receptacle for the core, whereas the outer barrel turns, driving the bit. These two types are screwed on to the end of a drill stem, which transmits the rotation of the machine to the 3m long core barrel. The latter must be hoisted when full, which takes some considerable time when the hole is deep. To avoid this, wire line core barrels have been invented.

Wire line core barrels resemble a lift in its shaft, and are made up of rods whose inside diameter is equal to the outside diameter of the core barrel. The inner tube of the core barrel is locked in the outer tube during drilling. When the sampling run is over, the inner tube is unlocked and returned by means of a hook on a thin metal cable which is raised by a fast working winch. This process is very quick and ensures the stability of the hole during recovery, but the cores are of a smaller diameter than those of the corresponding ordinary core barrel and it is more costly to purchase and operate (thicker bits).

Drilling is carried out with bits of different diameters, grouped in series of descending order. The most commonly used are the metric or 'Swedish' series made up of 9 successive diameters and the American Diamond Core Drill Manufacturers' Association (ADCDMA) series. Four of the most commonly used diameters from the ADCDMA series are:

Designation	Diameter of hole	Diameter of wire line core
H	96mm	64mm
N	76mm	47.5mm
B	60mm	36.5mm
A	45mm	27mm

A second letter is commonly appended to the letter codes shown, indicating the core barrel series. The most familiar is the X series (AX, etc.); a W (or Q) wire line series is also used.

1.2.2.3 Machines

Coring drills consist essentially of a swivel head with a hollow spindle, into which pass the rods which are rotated by the spindle using a screw or hydraulic jaw chunk; the pressure on the bit is applied in most cases by hydraulic jacks. The drills are equipped with a working winch for raising and lowering the rods.

Every manufacturer has a range of machines. The smallest machines include the Winkie by Anit, the Prosper by Atlas, and the X-Ray by Boyles Bros, which can be backpacked. Their performances are limited in depth to 10 to 50m depending on the rock, in diameter, and in application (relatively stable rock).

Middle range machines are the most numerous. They can drill to depths of 100 to 500m, the largest of them being able to reach 1000m: Longyear, Boyles Bros, Atlas Craelius, Acker, Sprague and Henwood and Joy are among the major manufacturers.

More and more manufacturers are producing machines with all- hydraulic functions such as the BE1 and BE2 from Bonne Esperance (France), the HC 150 from Longyear (USA), the B0, B1 and B2 from Wirth (Germany), the JKS 300 from Smit (Canada) which also has the advantage of being light, and the automatic Diamec 1000 with high rotation speed from Atlas (Sweden).

Machines also exist that are designed for doing both percussion and rotation work, such as the VPRH Foraco and Person Air Drill (survey type).

Coring drills come in a wide variety. Their efficiency basically depends on the rock drilled. For rocks of average hardness and quality, an advance of 6 to 8m per shift, including moving and setting up time, is normal for holes of 200 to 300m in depth. The cost can vary from \$100/m to \$300/m.

2 Sample recovery

2.1 Destructive drilling: recovery of cuttings

Destructive drilling is carried out for rough calculation and to provide an answer to the question 'Is there mineralization or not?'. It is used at the stages of the exploration or prospects and the study of deposits. Examination of the cuttings will reveal the nature of mineralization and its approximate grade. It is therefore important that recovery is good and that the origin of the cuttings recovered is known as precisely as possible.

When using wagon drills or down-the-hole hammer drills the cuttings forced up by compressed air return to the surface between the drill stem and the edges of the bore hole, and are recovered at the exit by an appropriate system. If the formations traversed contain water, the water is forced up by the compressed air together with the cuttings.

2.1.1 INSTALLING THE EQUIPMENT

The drilling rig is made up of a drill and a compressor connected by a reinforced flexible hose. The drill, mounted on a truck with tracks, wheels or skids depending on the ground, is set up at the drill hole site. The compressor, which feeds the drill with compressed air, can be placed up to 100m from it so that the compressor will need to be moved less frequently in the course of the operation. However, it is essential that the compressor supplies the drill with about 12,000 litres/min (between 7000 and 15,000 litres/min) at pressure of 8 to 12kg/cm^2, so that the pressure on the hammer remains high enough to give a normal advance. It must be remembered that in ground containing water the pressure of the compressed air for raising the cuttings is reduced at the bottom of the hole by 1kg/cm^2 for every 10m of water above it.

Once the hole has been started, the first few metres, when drilled in superficialy friable ground, are often cased. Casing is not generally done further down, unless the hole traverses very fractured or karstic zones, which cause high losses of returning air and consequently, of cuttings. The 'Saturn' process makes simultaneous down-the-hole drilling and casing possible thus giving protection as and when unstable zones are intersected.

2.1.2 ROLE OF THE PROSPECTOR

The prospector responsible for carrying out a destructive drilling campaign has two main responsibilities. The first is to supervise the drilling and note any information which may be of interest from a geological and mining point of view. The second is to recover the most representative samples possible from the rock and ore drilled.

2.1.2.1 Technical supervision of drilling

Drilling is the responsibility of the head driller, who informs the prospector of any problems which may arise.

The necessity for accurate sampling imposes certain constraints on the driller which it will be up to the prospector to explain, and if the need arises, to enforce. The following points are essential for a good standard of drilling:

1) be sure that the machine is technically adapted to the drilling programme;
2) check the setting up;
3) make sure that the sampling system is properly in place before drilling starts;
4) take care that the sampling runs are maintained;
5) finally, and most importantly, be sure that the bore hole is thoroughly cleaned out between each sampling run. Insist on this to the driller; ask, even if it takes a long time, sometimes as much as 5 min, that the hole be blow cleaned while the drill stem is raised and lowered to a height at least equal to that of the length of the run, until the air coming out of the hole is practically free of cuttings.

The drilling sheet **(Table 6.1)** shows all the information that the driller should give: drilling time, pressure on the bit, wear on equipment, sizes of rods and bits, etc.

Drilling should not normally be stopped during sampling, except between two runs.

2.1.2.2 Recovery of cuttings

This is the main part of the prospector's work. The method and equipment described below result from the experience of the BRGM over many years **(Fig. 6.1)**.

Bore hole No:						Site:						Type:
Date:						Shift foreman:						Machine:
Bored		Length	Duration	Bits		Casing		Air pressure kg/cm²	Water	Rock	Recovery %	Observations

Table 6.1 *Example of percussion drilling form (To be filled in by the driller).*

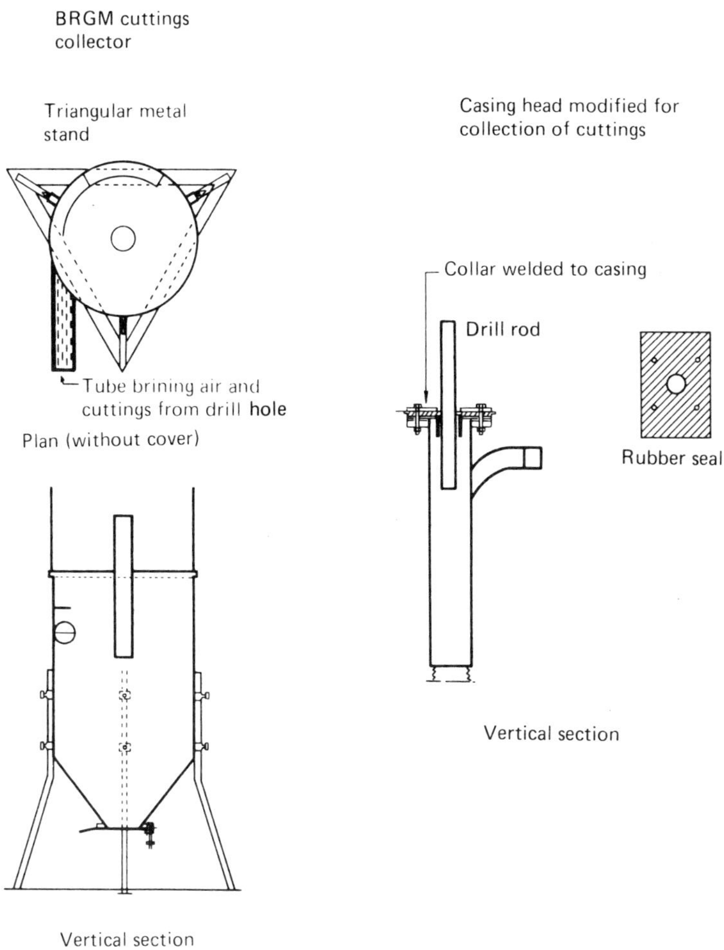

Fig. 6.1. *Cuttings collector for percussion drilling.*

Equipment necessary
The essential apparatus is the sludge collector, which should have a volume of at least 10 times that of the sample (which is often 20 litres).

This type of apparatus is frequently constructed on site. The details are given in Fig. 6.1. The cuttings and air arrive in the upper part of the cylinder. The cyclone created by the tangential arrival of the air is controlled by a thin arcuate inner plate whose friction slows down the incoming material which then accumulates in the bottom of the collector. The extraction of the dry or water sodden air is by a chimney in the form of a pipe opening below the inlet level of the cyclone.

A large diameter semi-transparent flexible hose links the casing head to the collector. Even if this hose sags under its own weight there is little risk of cuttings accumulating in the bend because of the high speed of the air carrying them.

However, if this were to happen, the hose could simply be propped up by a few stakes.

The rest of the equipment comprises:

1) two tanks, each big enough to hold a whole sample, even when wet;
2) a rubber mallet, to remove wet sludge stuck to the sides of the collector;
3) a pyramid shaped sample-divider for quartering samples which are dry and free from any grease that may, in the case of the wagon drill, have come from the drill rod coupling thread;
4) four tanks for the recovery of the quartered material;
5) a sheet of metal for the quartering of the wet samples, ('cake' technique);
6) a 1, 2 or 3mm screen according to the availability;
7) one or two 8 litre pans;
8) a small wash pan for examining coarse cuttings;
9) two trowels;
10) in dry areas a jerrycan of water for washing the coarse cuttings;
11) a basin or tank for the washing;
12) a shovel and a rake to prepare the storage area for the piles of residue;
13) canvas and plastic sample bags (3 litre);
14) Caubere sample tubes with labels to go in them;
15) a large bag not necessarily new to stock the sample to be washed in the pan;
16) a small brush;
17) string or twist-on metal fasteners used to seal bags;
18) bottle of hydrochloric acid diluted to N/10 and the normal reagents;

if necessary:

19) a magnet;
20) a dust trap.

Personnel
At every site, and during all drilling procedures, a geologist's assistant should follow the drilling and take samples. He will be helped by three or four men, one of whom works on the small wash pan. The geologist's assistant is in charge of the above equipment and, under the supervision of the prospector, carries out the sampling as follows:

1) one man on the cyclone;
2) one very experienced panner who can identify minerals, probably the geologist's assistant;
3) one man quartering;
4) one man on labelling, tying and sorting the various bags;
5) a senior experienced prospector coordinating the work and taking notes.

During drilling, workers 1 and 3 will be fully occupied supervising the sample-divider for the liquid samples.

Sample recovery

The length of drilling per sample run is determined in advance by the geologist, according to the nature of the ground. As a general rule, recovery is carried out every time another rod is added, usually every 2.40 or 3m outside mineralized zones or sometimes even in them when the thickness of the deposit is known, and every 1m in ore. More frequent recovery may sometimes be desirable. The theoretical volume of the sample to be recovered is calculated using the diameter of the hole and the length of the run. For a 1m run it varies from 1.25 litres for a small wagon drill to 22.7 for a large down-the-hole hammer drill.

During the drilling of the run the geologist's assistant makes observations in the columns of a notebook such as that shown in **Table 6.2**, noting in particular the colour of the dust from the hole, which he sees at the chimney on the cuttings collector, the exact depth at which the colour changes, and any other incident or observation which may be of use, such as the wetness of the sample.

When the cuttings, sometimes called sludge by the driller, are wet they stick to the walls of the hole, to the casing lining it and to the rods. It is then necessary to circulate water in order to clean the hole and expel the sludge. If the inflow of underground water is sufficiently large, this will keep the hole clean. In this case, the recovery depends on the quantity of ascending water. If it is not too great, not more than 100 litres/m, then recovery is possible and it is simply a case of sending air, water and sludge returns to a tank (a 200 litre half-barrel will do nicely) which will serve as a decompression or tap swirl chamber, and letting it run into a tub or riffle box to recover each sample. The collector in Fig. 6.1 may also be used by opening the recovery hatch slightly, taking care to place a tub below the hatch. It must be remembered that much of the fines escape and that the qualitative representation is very distorted.

At the end of the run, when the machine stops, all the sample is drawn out of the collector into one of the tanks set aside for this purpose, and it is allowed to decant for several minutes if necessary. Once the collector is emptied of all contents including the fines removed from the walls with the mallet, the driller begins cleaning the hole as mentioned previously. The air blown in the course of this cleaning leaves via the collector with the hatch open. The hole and the equipment are thus clean for the next run.

Whether or not quartering is done will depend on the quantity of samples collected. Normally, those furnished by the small wagon drills are all put into bags with the following information (in indelible ink) on the bag and on a thin wooden or plastic block placed inside the bag:

1) site name, drill number;
2) sample number;
3) start and end of run depths, length of the run. Bags, labels and blocks are prepared before drilling starts with the information on points 1 and 2. Number 3 is added during drilling.

When quartering has been done, especially in the case of down- the-hole drilling, the sample, if dry, is passed over the pyramidal sample-divider, Jones sample-divider or Minemet riffles box as many times as are necessary to obtain three samples. Two samples of around 1.5 to 2 litres should each be packed in a double lined bag, one

Drill hole No Date					Site					Geologist	
Drilled from to	Length of run	Colour	Water	% recovery	Panned yes/no	Tests		% ore minerals in rock	% and description cuttings	Description of ore minerals	Miscellaneous
						HCl – or + or ++	Mag –or+				

Table 6.2. *An example of the sample log (percussion drilling) (to be filled in by the geologist's assistant).*

plastic and one canvas, and numbered as before. One will be sent to the laboratory for analysis and the duplicate is kept at the camp.

One 3 to 4 litre sample should be treated as follows and noted on the drilling sheet and on the log:

1) After screening at 1, 2 or 3mm depending on the screen available, the screen oversize should be examined by hand lens for each drill hole section, while the screen undersize should be collected in a bag, then washed in the pan, for examination of the concentrate under the binocular microscope.
2) Equipment is carefully cleaned and the drilling sheet is filled in after any necessary acid and magnet tests, etc.
3) Finally the bagged samples are sent to the laboratory. These operations are carried out in their entirety for each drilling operation before proceeding to the next operation.

In the case of wet samples, for the down-the-hole hammer drill and in all cases for the wagon drill because of the presence of grease the cuttings are quartered by the cake method on a metal sheet free of the metal being drilled for until the same samples are obtained as before.

What remains from the various quarterings is set down in rows of 10 piles, in rows always arranged in the same way (first run in top left and first heap from each batch of 10 at top, for example), on a piece of ground prepared for this purpose with a rake and shovel.

Sample representiveness

Destructive drilling results in qualitative observations but quantitative estimation of mineralization is much less sure than for core drilling.

Errors and approximations are caused by loss of sample, especially fine fractions during drilling when, for example, the rock is fractured; by dilution or concentration of the ore in the samples; and by incorporation of material dislodged from the walls, all of which are inherent in the drilling method. Lack of care or precautions during drilling, in sampling and analysing could also be added, but these faults can and should be remedied.

Sample loss in drilling can easily be discovered by estimating the percentage of recovery, ie comparison of volume drilled and volume recovered, taking account of the expansion coefficient of the cuttings is normally 2/3. This loss, varying between 5 and 100% depending on rock conditions will always occur. In the case of a total or almost total loss, it is preferable to stop drilling.

The loss of fines varies according to whether the cuttings are dry or wet. When dry the losses seem only to occur in sample recovery, where the fines are easily carried along by escaping air. The creation of fines in drilling varies according to the hardness and fracturing of the rock (resistance to breaking); according to the striking force of the bit and the rate of advance, and thus according to the blow cleaning time. The estimated loss varies from 5 to 30% in volume.

For a wet sample the losses are much greater and will occur at two points:

1) during recovery in the sampler where the emulsified water contained in the escaping air takes with it much fines; even in the case where the sludge is

recovered in decanters, the fines are to a large extent carried away in the water and lost;
2) during decantation flotation of the sulphides may occur essentially due to the presence of grease or oil and their being carried along by the water.

Losses are generally greater than 20% in volume.

Cleaning of the drill stem and of the sampling and quartering apparatus between sample collection limits contamination by tailings (dilution) or by ores (concentration). Contamination may occur due to incorporation of material dislodged from the walls, especially soft rocks which have been swept along by the rising air, but this is not important and does not affect sampling.

Finally, the greatest source of error is in sample loss. The percentage lost can be estimated but its nature is more difficult to estimate. This can be remedied in part by the use of extreme care in all cuttings recovery operations and in perfectly dry ground by the use of dust traps (eg Atlas Copco DCT type). These contain bags of synthetic material with an increasingly fine mesh for filtering the air loaded with dust and recovering it even when the particles are very small. Each bag in the trap should be removed after every run, numbered and sent to the laboratory to have its contents analysed. A new set of bags is placed in the trap for the next run. This procedure is costly, so it is often preferred to empty the trap bags by shaking them into ordinary plastic bags, and to put the whole set back in the machine. But the risks of contaminating the cuttings of one run with those of the one before are high. This expensive technique, which makes it possible to recover the finest particles, can only be applied to perfectly dry rock which is rare.

The organization of a percussion drilling site is summarized in **Table 6.3**.

2.1.2.3 Making out the drilling log

All of the above work should result in an accurate document bringing together information on working conditions, geology and mineralization.

Table 6.4 is a combination of observations and interpretations. However, the separation of these two categories of information is always possible. All documents should be carefully filed away.

2.2 Core drilling: core recovery

The core faithfully reflects the rock traversed: its complete recovery is therefore very important.

First it is advisable, as far as possible, to use a drill crew with experience in drilling for minerals, the main concern of which is good core recovery and to help them by giving them precise instructions.

Request that drilling in ground known to be bad be started with a coring drill of the maximum diameter possible according to the depth of hole concerned, in order to be able to drill as far as possible using progressively decreasing diameters downward. The larger the diameter of the core the greater the chance of a good recovery. Other factors which come into play when a maximum recovery is desired in difficult rock are the weight on the bit, the speed of rotation, the rate of flow and nature of the circulating fluid, but these are more the concern of the driller.

	Work done during advance	Work done within 1/2 to 1 day of advance	
Drilling sheet	→ Drilling sheet (geologist's assistant)	→	Technical data, Mechanical tests
Cuttings recovery per 3m (10 feet) or 2.4m	→ Quartering – Duplicate sample, 1.5kg stored on site		
	– Lab sample **1.5 kg**	→	Cu, Pb, Zn, Au, Ag analytical results
	Remainder; estimation of overall mineralization	→	Estimated % mineralization
	Screening, Ø 1, 2 or 3mm → oversize	Petrographic determination on coarse cuttings →	Lithology
	Undersize, possible panned, heavy mineral concentrate →	Binocular microscope examination, mineral determination →	Mineralization
			Drawing of log
Drilling crew	1 geologist's assistant, 3 or 4 labourers	1 geologist, 1 prospector	

Table 6.3. *The organization of a percussion drilling site.*

BRGM department, section ....

Drill number ....
Drill hole specifications ....
Depth of drill hole ....
Coordinates X = ... Y = ... Z = ...

Drill hole progress chart

Site or locality name ....
Percussion drilling ....
Section at (scale) ....
Filled in by ....

Direction		Inclination				Started on			Finished on			By (driller(s))		
Technical data						Macroscopic examination			Analyses					
Depths	Bit diameter	Drilling time (min) 15 30 45	Com-pressed air pressure	Depth of water-table	% recovery cuttings 0 50 100	Depths	Rock type log	Minerals	Barren panning metals found (choose approp-riate symbol)	Results (indicate if given in ppm (g/t) or %) (in metal per column)				

Table 6.4. *A percussion drilling log.*

Request that the length drilled per run when approaching, and when in the mineralized zone, be reduced even to 1 or 0.5m if necessary. This has the effect of reducing the risk of losing core by wear. In fractured ground a piece of rock may wedge itself in the coring drill and wear away all or part of what is drilled afterwards, and the larger and harder fragments may reduce the smaller and more fragile ones to dust. The reduction in length of run also makes it possible to locate more precisely possible losses when the ore is being traversed. If, despite this, loss still occurs, the driller can be asked to collect at the drill hole outlet the sludge brought up by the circulating fluid during the sampling run.

The sludge is returned to the surface by an injection fluid which may be water or mud, depending on the nature of the ground to be cored. In stable formations water will usually be sufficient to cool the bit and return the fragments produced by the action of the grinding surface of the bit on the rock. In bad ground it is recommended to work with a light mud of water with clay or bentonite, which will ensure the return of the coarsest fragments and leave a thin film of solid particles (cake) on the walls of the bore hole which seals off minor circulation losses and reinforces the walls. The speed of ascent of the sludge, which should be between 45 and 70cm/s, depends essentially on the rate of flow into the section of the annular space.

Cores thus recovered are taken directly from the core barrel and placed in the core box where they are laid out from left to right beginning with the uppermost piece of core (start of run).

2.2.1 ARRANGEMENT OF THE CORES

On each box the direction of the cores is indicated by writing 'top' and 'bottom' on the edge of the planks forming the framework **(Fig. 6.2)**. The cores from each run

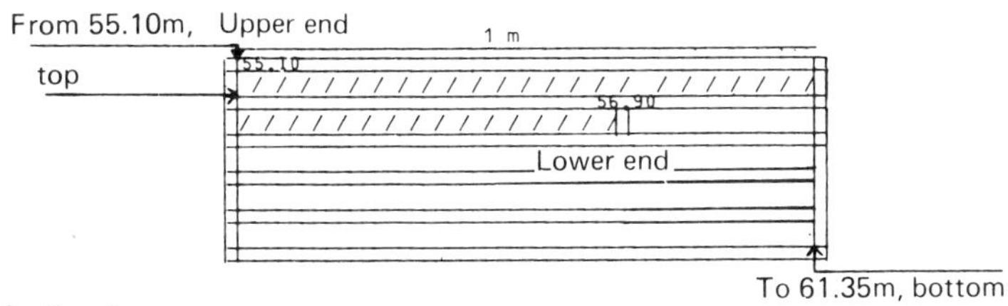

Fig. 6.2. *Core box.*

are separated from one another by a cube shaped piece of wood, on which the following information is written:

1) start of run depth;
2) end of run depth;
3) distance cored;
4) length recovered.

The point where the run ended should also be shown on the edge of the separating lath at the lower end of the core. It is advisable to punch this information on a metal tag which is then nailed onto the wooden peg and the separating lath. A 1cm wide strip would be very suitable. Before closing the box, write the following on one side which will remain visible after storage:

1) name of site;
2) drill hole number;
3) box number;
4) starting and finishing depths of distance cored (eg from 55.10 to 61.35m).

2.2.2 CORE SAMPLING

After the first examination, which makes it possible to draw up a log and locate any mineralization, the pieces of mineralized core are cut in half lengthwise. One half is kept as a duplicate, while the other is usually cut again and sent off for detailed laboratory examination.

There are two devices that the prospector can use to split the core. One is the old fashioned and heavy core splitter, rarely used nowadays, which breaks the core very irregularly, but can still be useful in certain circumstances. The other tool is the core saw whose disk can cut even the hardest pieces of core neatly. The disk is cooled by water during cutting and is powered by an electric motor or internal combustion engine.

2.2.3 EXAMINATION OF THE CORES: DRAWING UP THE LOG

After having set out the cores in the box a start is made by measuring the length of the core recovered per run, the relation between core length and length of run giving the recovery rate. This should be done as soon as possible after the extraction of the core, so that in the case of poor recovery the driller can be asked as quickly as possible to take the necessary steps.

Then follow the petrographic, mineralogical and structural examinations of each run. In general, all the information is noted in a notebook before it is put on the final document, the drilling log, and samples are taken for analysis. As far as structures are concerned remember that the only possible measurement of a plane is that of its inclination in relation to the drilling axis. The direction and true dip of such planes, with the exception of those perpendicular to the drilling axis, cannot be determined. Only by using core orientation devices can the direction and dip of structures be measured. These will be mentioned at a later stage.

The following drilling log is meant to be as complete as possible; but other columns may be added as needed, eg calculation of the weighed average grade, found by regrouping different mineralized sections of the bore hole.

The inclination of the bore hole is always given in relation to the horizontal, in drilling in galleries the inclination figure is always preceded by a + sign for upward bore holes, and by a – sign for downward bore holes. In the 'drift' column a horizontal line is drawn where the measurement was made above this line the measured direction is noted and below it the measured inclination. 'Not measured' is written in the column if no measurement has been made **(Table 6.5)**.

2.3 *Controlling the inclination and direction of bore holes: core orientation*

It is always useful, and sometimes vital, to know the trajectory followed by the bore hole in relation to the original axis, whether the bore hole is inclined or vertical. This is the case, for example, where the intersection hoped for when drilling for a vein is not obtained; this may be due either to the absence of the vein or to drift of the bore hole which has missed its target.

2.3.1 INSTRUMENTS

Several types of orientation instruments are on the market. Three of them are described below: the HF clinometer, the Tro-Pari and the DDI Fotobor reflex camera.

2.3.1.1 HF clinometer

The directional drift is not measured with the clinometer, the successive inclinations being assumed to occur in the azimuth plane at the origin of the bore hole. The measuring principle is based on the glass etching property of hydrofluoric acid (Appendix 1 for uses). This method is rarely used nowadays.

2.3.1.2 Tro-Pari

This is the most commonly used instrument. It measures the direction and the inclination of any bore hole, whether up or down. It consists of a hemispherical case mounted on gimbals. Its equatorial plane is occupied by a compass and the pole by an opening through which a pawl clamps on to a semicircular graduated toothed rail. The hemisphere contains a clockwork mechanism regulated by a timer.

The time delay device is set so that the compass locking wedge and the pawl of the clinometer both lock after a lapse of time calculated generously enough to allow for the lowering of the instrument to the required depth in the hole, which it should reach at least 10 min before the clamping time, and remain for 5 min afterwards. The instrument is then placed in a special bronze cylinder with an outside diameter approximately equal to that of the drill rods. Make certain that the arrow marked on the instrument is pointed towards the inside of the cylinder, ie towards the bottom of the bore hole, and that the pole of the hemisphere is well above the toothed rail. In spite of these precautions the pole sometimes, though rarely, moves back during the operation, which then has to be repeated. The brass cylinder is screwed on to three aluminium rods to avoid any effect on the compass from the normal steel rods; then the number of ordinary rods needed to reach the required depth for them are added. When the set time plus at least 5 min has elapsed, the instrument is raised, taken out of its cylindrical case, and checked to make sure that the arrow was in fact pointed towards the bottom of the hole, and the measurements on the compass and on the graduated toothed rail are read off in degrees. Should the arrow head by mistake have been placed upside down, the measurements shown on the compass would have to be corrected by 180°.

More detailed operating instructions come with the instrument.

2.3.1.3 ABEM DDI Fotobor reflex camera

As it uses a compass for measuring directions, the Tro-Pari can only be used in non-

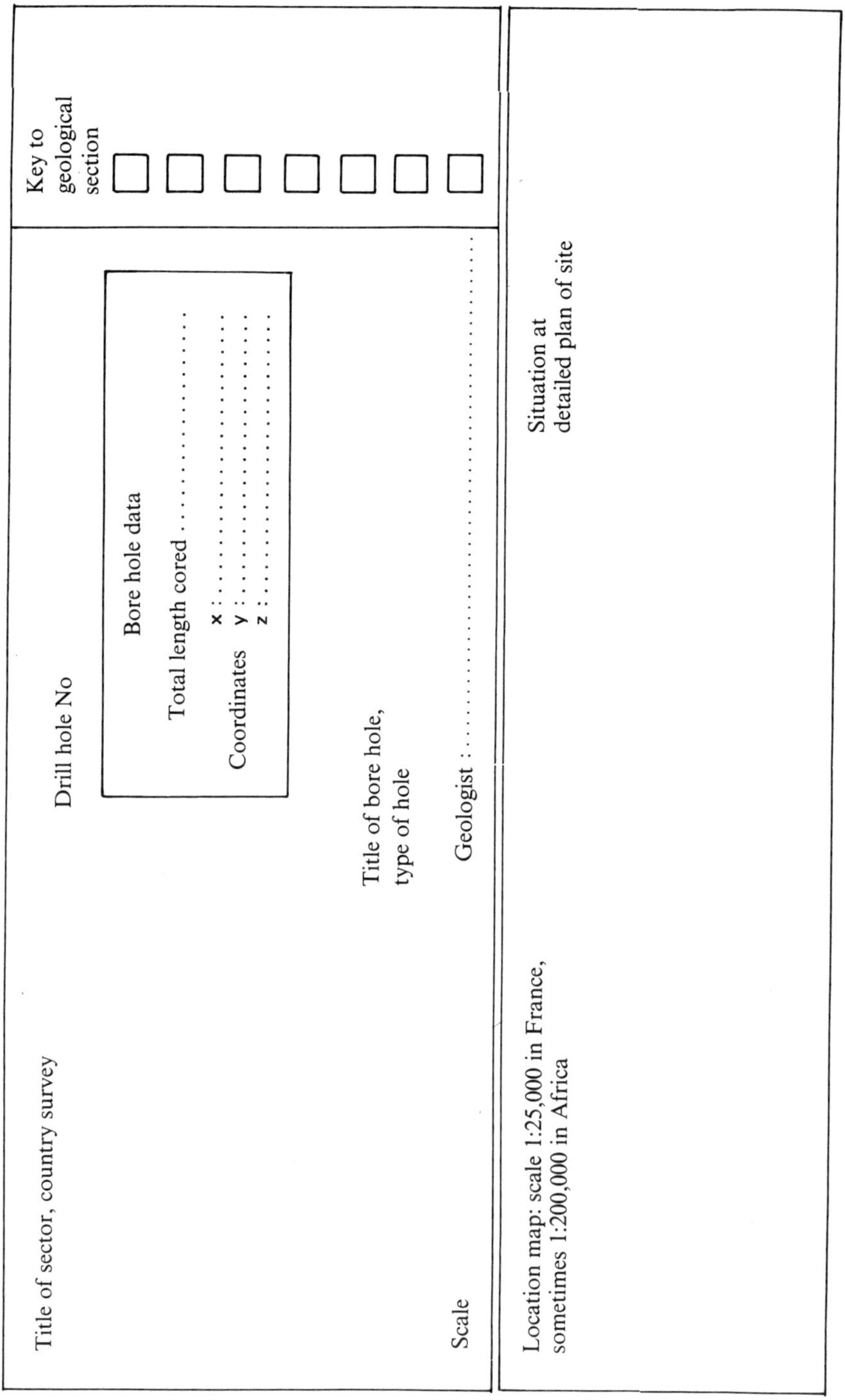
Title of sector, country survey
Drill hole No
Bore hole data
Total length cored ....
Coordinates x : ....
y : ....
z : ....
Key to geological section
Title of bore hole, type of hole
Scale
Geologist : ....
Location map: scale 1:25,000 in France, sometimes 1:200,000 in Africa
Situation at detailed plan of site

Direction							Inclination				Begun on		Completed on			Crew	
	Coring			Depth			Section	Structural data			Macroscopic examination		Analyses				
Storage box No	Diameter	Percentage	Recovery	True depth	Length	Deviation		Ts. drill hole axis	Ss. drill hole axis	Calculated structural data	Petrography	Mineralogy	Polished section	Thin section	Cut-off	Ore, ppm % Ore	Ore, g/t

Table 6.5. *A core log.*

magnetic rock. Electronic devices have been developed for magnetic rocks but they are complicated to use. The most attractive is the DDI Fotobor reflex camera made by ABEM which comprises a long flexible pipe that adapts itself to the curvature of the drill-hole, on which are marked at 3m intervals, three reflective rings, one at the bottom, one 3m up, and one 6m up. A multi-shot camera photographs these three disks, the relative position of which recorded on a special background, shows the progressive drifting and by calculation, the coordinates of the measuring points. Finally, gyroscopic processes now being investigated should become available soon.

2.3.2 PLOTTING THE MEASURED DIRECTIONS AND INCLINATIONS

In practice, changes of course in bore holes, in both inclination and direction, are progressive, and very sharp normal changes in orientation are extremely rare. Every deviation measurement gives the direction and inclination of the tangent of the drill trajectory curve at the point where they were measured. For the section, accordingly, the length of the bore hole covered by the measurement is equally divided above and below the measurement made.

For example, in the case of one measurement every 40m the first measurement at 40m gives the direction and inclination of the drilling zone from 20 to 60m, the second measurement at 80m gives the direction and inclination of the drilling zone from 60 to 100m, and so on. Obviously the graphic representation will improve with increasing frequency of the measurements. In most cases one measurement every 40 to 50m will be sufficient to obtain a very satisfactory graphic representation, but more frequent measurements (every 20 to 30m for example) should be used for highly deflected bore holes **(Fig. 6.3)**.

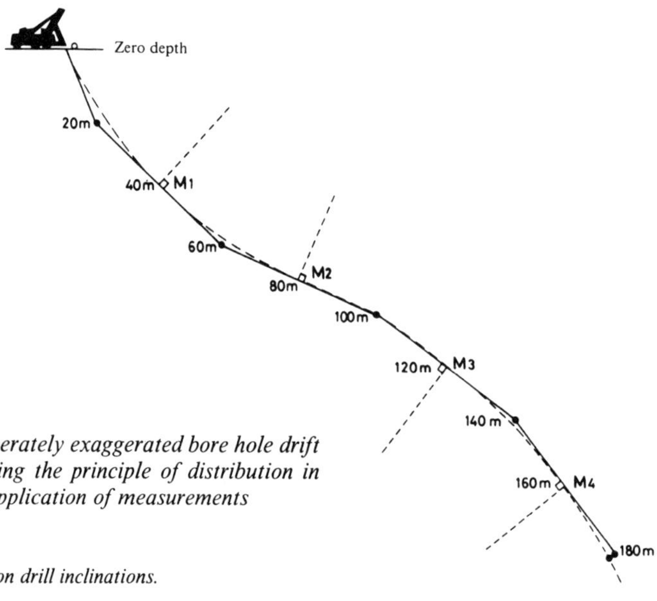

Deliberately exaggerated bore hole drift showing the principle of distribution in the application of measurements

Fig. 6.3. *Section drill inclinations.*

2.3.2.1 Method of graphic representation

All new points measured can be calculated by trigonometry. Graphic representation is most useful because it is immediate. Plotting in the two planes (horizontal and vertical) is done simultaneously, and an example is shown in Fig. 6.4.

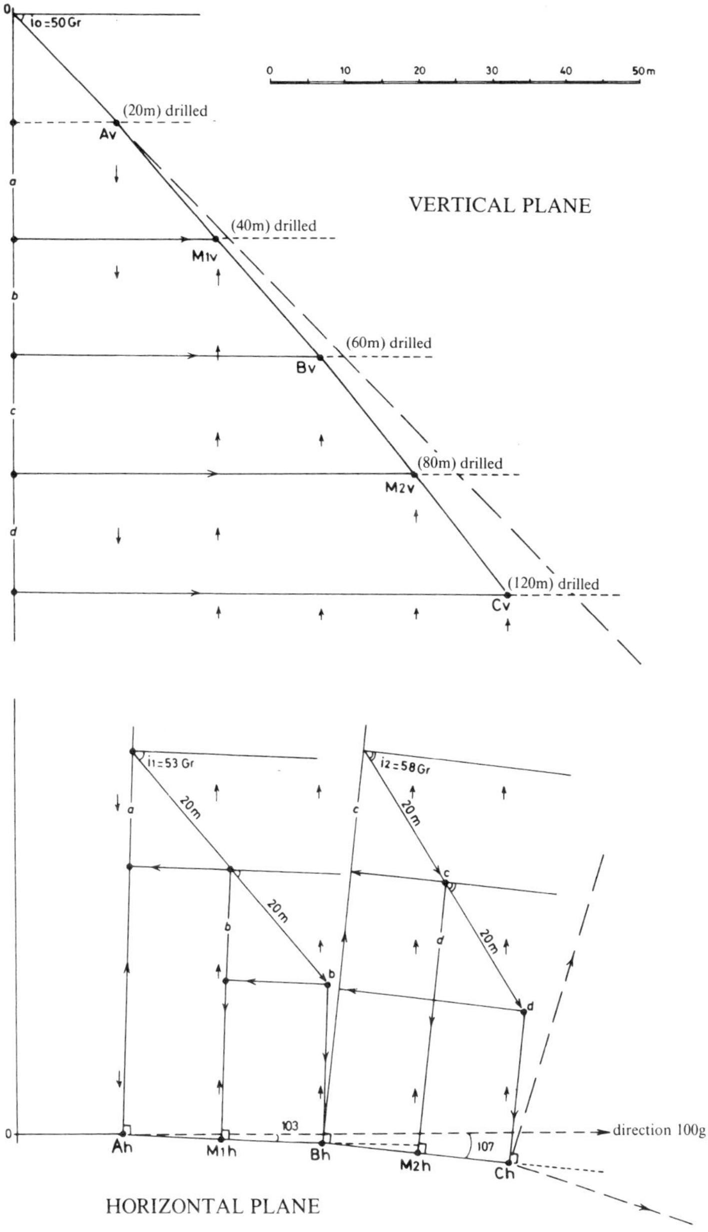

Fig. 6.4 *Example of graphic construction of bore hole deviation.*

On a sheet of millimetric graph paper the vertical plane (section) is drawn in the upper part and the horizontal plane (plan) in the lower part, on which will be represented the inclinations and the directions measured, respectively. Each plane represented passes through the origin of the bore hole. Remember to take into account the space necessary to accommodate the depth of the bore hole between the section and the plan. For medium depth bore holes (200 to 400m) a scale of 1:500 is useful.

A point by point description of the procedure follows.

2.3.2.2 In case of unevenly spaced measurements

Measurement M1 at 30m
M2 at 60m
M3 at 100m
M4 at 150m

The intermediary points will therefore be:

A at 15m, 1/2 distance from origin to M1
B at 45m, 1/2 distance from M1 to M2
C at 80m, 1/2 distance from M2 to M3
D at 125m, 1/2 distance from M3 to M4

2.3.2.3 Drawing-up the section

Angular measurements in this example are given in 'grades', using the system of 400 divisions (grades) in a circle. The grades system is most popular in France but it is a good example.

Initial direction of the bore hole: 100 grades, inclination: 50 grades.

Measurements taken at:

40m M1 giving direction 103 grades; inclination i1 53 grades
80m M2 direction 107 grades; inclination i2 58 grades
120m M3 direction 120 grades; inclination i3 60 grades

The equidistant points are A, between 0 and M1, B between M1 and M2 and C between M2 and M3.

For greater clarity the letter v has been added after every point on the vertical plane, and the letter h after every point on the horizontal plane.

Let 0 be the origin of the bore hole. The first part of the bore hole (0 to 20m) is not covered by the measurement M1, and is drawn directly on to the vertical plane (VP). OAv=20m is plotted with i0=50 grades. By projecting downwards the point Ah on the horizontal plane is obtained.

From now on, all constructions will be made starting from the horizontal plane (HP).

Starting from the point Ah, the section covered by M1 is plotted. A straight line of direction 103 grades is drawn from Ah. A perpendicular is raised from this line at the point Ah. From any point on this perpendicular the representation of the inclination of this bore hole at the point M1 is drawn (let i1=53 grades) and this length drilled from A to M1, ie 20m is plotted. The point 'a' is obtained, which by vertical projection gives the point M1h and by horizontal projection gives the vertical depth reached (length 'a').

The length 'a' is plotted on to the VP from the vertical depth of the point Av and gives the vertical depth of the still unknown point M1v. From this point a line is drawn horizontally to the right. M1v will be the point of intersection of this last straight line with the position 'in space' of the point M1h, of the HP raised vertically along the millimetre scale as indicated by the small arrows.

We return to the HP to locate the point 'b' at 20m from 'a', still following the same inclination (53 grades). The point Bh is obtained by the vertical projection of 'b' on to the line M1h-Bh, of direction 103 grades. The horizontal projection of 'b' gives the vertical depth (distance 'b') which is plotted beneath the vertical depth of M1v giving the vertical depth of Bv, which is obtained by the intersection of which a line is drawn horizontally from this point.

Moving on to the section covered by the measurement M2, a straight line of direction 107 grades is drawn from Bh. At the point Bh a perpendicular is raised to this line, a straight line of new inclination (58 grades) is constructed on which is plotted the 20m drilled between B and M2, and so on until the end of the drilling.

So in fact it is a case of constructing the horizontal deflection (from the direction 100 grades) of the vertical section of the bore hole viewed from above the horizontal plane.

The line O, Av, M1v, Bv, M2v, Cv reproduces as accurately as possible the vertical section of the bore hole on its original vertical plane. The line O, Ah, M1h, Bh, M2h and Ch represents as accurately as possible the horizontal section of the bore hole on its original horizontal plane.

2.3.3 CORE ORIENTATION

Although the processes described above make it possible to locate the bore hole perfectly in three dimensions, they give no information as to the position occupied by the core in the rock before being extracted. Therefore it is impossible to reorient in space and thus to measure in relation to a fixed point (the North for example) the structural elements measured on a core in relation to the drilling axis, such as strike and dip of beds, fractures, veins, etc. The interpretation of a drilling programme can only ever be partial and raises problems which at best one can only attempt to solve by drilling more bore holes.

The value of the information provided by the orientation of cores has not been ignored by manufacturers, who have, however, produced devices which so far have proved awkward to use and often complicated and fragile.

The BRGM structural geologists and drill engineers have studied the problem and have come up with an accurate, simple and robust device which does not affect the content of the cores, since it is totally independent of them: the BTV 20° to 160° core sample orientator which is patented by the BRGM.

2.3.4 BTV 20° to 160° CORE SAMPLE ORIENTATOR

It is adapted to the wire-line drilling technique and works in bore holes with an inclination of between + 70° (70° upwards) and – 70° (70° downwards). Its use may be entrusted to the drill crew and does not in any way modify the drilling conditions.

Lowered between every run instead of the inner tube the BTV takes an impression of the bottom of the hole using sliding feelers which release a weight mounted on an axis. This preliminary operation makes it possible to mark the lower face of the core after it is removed from the core barrel.

The bore hole itself being regularly surveyed by conventional methods, the geologist can then measure the orientation of all the structural elements observed.

The BTV 20° to 160° core sample orientator is made with AQ, BQ and NQ diameters, with adaptations for larger diameters. The standard equipment is composed of:

2 orientator blocks with feelers;
1 filter fitting;
1 sealing fitting;
1 adaptor pipe (can be varied according to drill equipment used);
1 inner core barrel head corresponding to the diameter and the drill equipment used;
2 feeler protection rings;
1 counter-impression block;
1 orientation bench;
Accessories: small tools, spare parts.

References

Jeanne, L; Lorand, M (1960) Note on the use of non-motorized drills by the BRGM in mineral prospecting in western Africa. BRGM Report, Dakar-Senegal.

Villeneuve, B (1977) Drilling in mineral exploration. *Ann. Mines, Fr. special issue* 'La prospection miniere'.

Appendix 1

Measurement of bore hole inclination by HF clinometer

Hydrofluoric acid principles

The inclination to the horizontal α is measured on the tube by the acute angle formed by the horizontal plane (indicated by the mark left by the acid on the tube) and the generatrix of the tube passing through the highest point or the lowest point of this mark **(Fig. 1)**.

Correction tables: because of capillarity (which varies according to the diameter and the acid strength) the angle read on the tube is α' (always greater than true α). Experimental correction tables exist for the conversion from α' to α **(Fig. 2)**.

Remarks, ACID, industrial strength of 48%. For shallow bore holes, dilute with 1 part of HF to 5 parts of water. *Pour the acid into the water*. For this proportion leave at the bottom for 30 min. Modify proportions and etching period with depth. TUBES. Use 'culture' tubes, the walls of which are very straight instead of test tubes. Do not use Pyrex since it does not etch well.

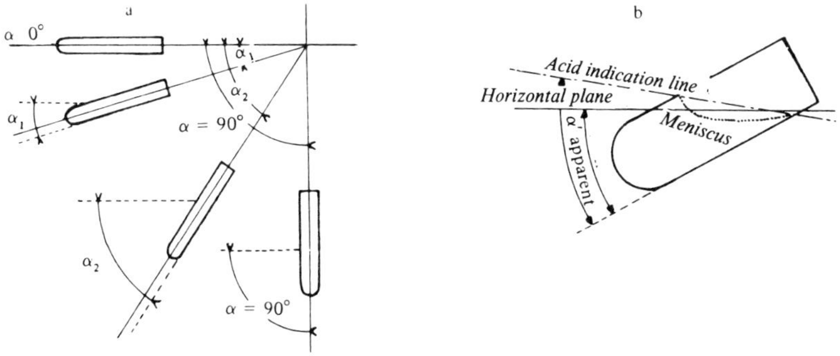

Fig. 1

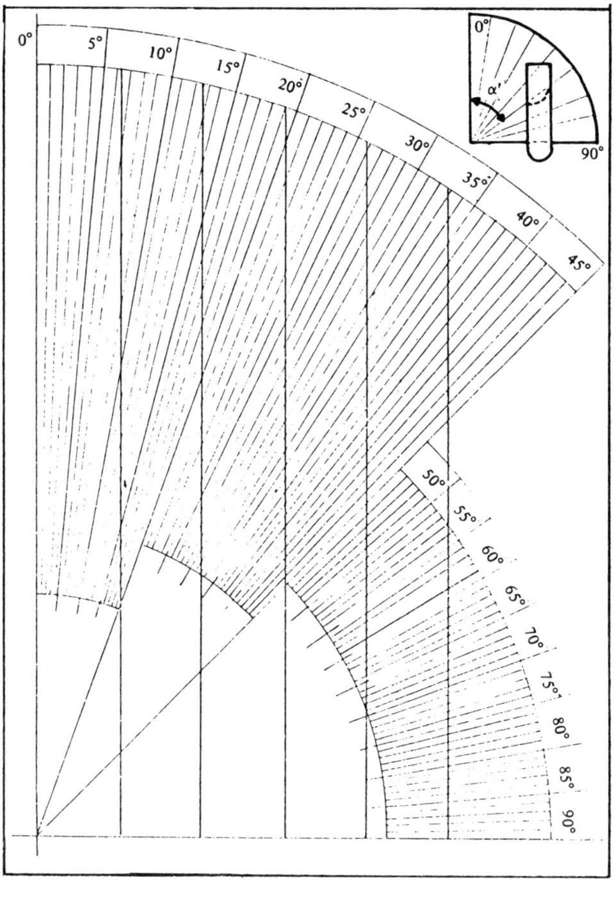

Fig. 2

Chapter 7

The role of the prospector in mining

Mining techniques draw on the talents of specialists, the mining engineers. However, a prospector may have to initiate certain minor exploration works in a gallery. This requires the knowledge of the methods of timbering, the use of explosives, ventilation and drainage. The information offered in this chapter will enable him to cope with these tasks. This chapter also deals with the different techniques for sampling in mining works and surveying galleries.

1 Small-scale mining works

1.1 Driving of galleries

The following terms are employed:

1) crosscut: a gallery which intersects rock beds up to the contact with the mineralized zone;
2) opening: a gallery opened in the mineralized zone along its length;
3) drifts: galleries driven on either side of the opening to determine the thickness of mineralization.

The following expressions are also employed:

1) the working face of the gallery is the point reached by the gallery at a given moment, before penetrating further;
2) sides or walls of the side surfaces of the gallery;
3) top or ceiling of the gallery;
4) floor of the gallery **(Fig. 7.1)**;
5) working or cutting, in loose formations, work can be carried out by pick or with a pneumatic pick hammer. The pick hammer is a striking instrument with a reciprocating motion in which the tool is a point.

Principle: under the action of compressed air, the piston periodically strikes the point as a hammer strikes a wedge. During operation the worker keeps the hammer against the mass by applying some pressure.

In hard formations, where working becomes difficult or even impossible with the pick hammer, mining is carried out by explosives. The explosive is inserted in shot holes laid out in accordance with a firing pattern and drilled by means of rock drills.

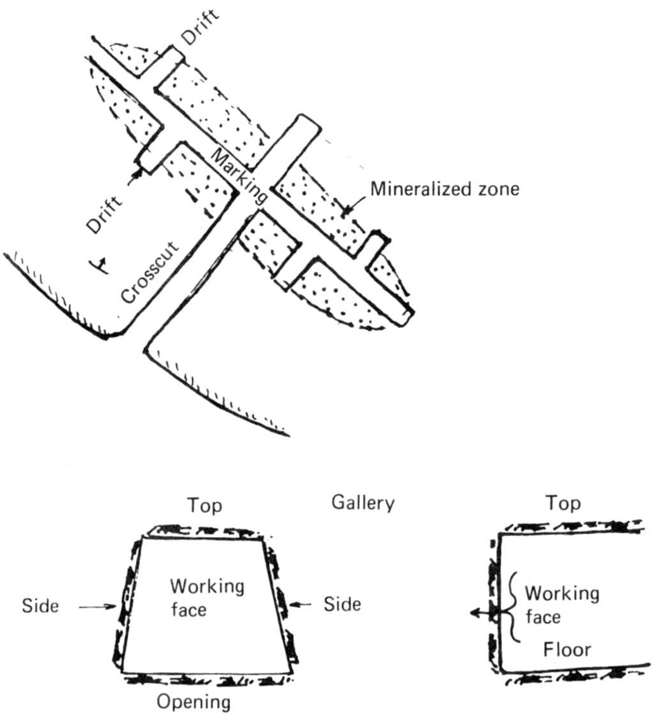

Fig. 7.1. *Mining terms.*

1.1.1 DRILLING OF SHOT HOLES

1.1.1.1 Rock drills

In the pneumatically driven 'drill' a piston in the forward stroke strikes a drill terminating in a cutting point. In its return stroke the piston rotates a certain amount enabling the tool to strike the entire surface of the hole successively. The rock is destroyed by crushing.

Most rock drills **(Fig. 7.2)** feature water injection. The water passes inside the hammer through a hollow needle and circulates inside the drill. This water cools the bit and removes the 'cuttings'. Water injection is mainly intended to eliminate dust and the risks of silicosis for the miners. The drills are quickly worn and must be sharpened frequently using a special grinding wheel.

1.1.2 FIRING PATTERNS

The firing pattern comprises the following:

1) the position and direction of the holes;
2) their number and charge;
3) choice of the type of detonation;
4) firing order.

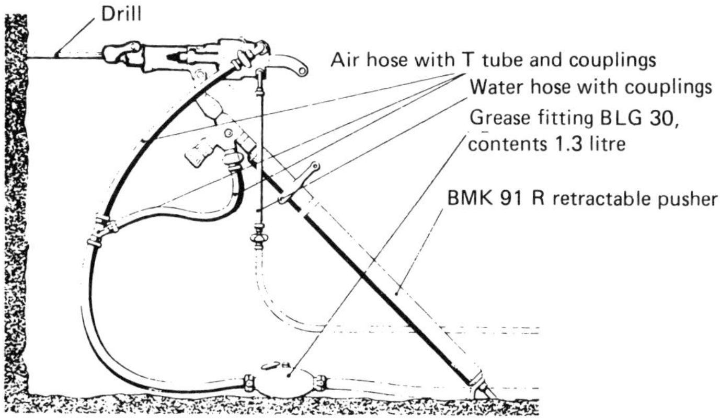

Fig. 7.2. *Rock drill.*

The choice of the firing pattern is also dictated by the type of rock encountered, the regulatory requirements and economic conditions (particle size distribution). To obtain a satisfactory yield from the explosives release surfaces must be created. For gallery driving this is the purpose of the 'plug', which is nothing other than a void created at the centre of the working face by a blast preceding that of the other holes.

The success of the plug blast is essential for the normal progress of shot firing. Several types of plug are employed, and the most widespread are:

1) the convergent or taper plug;
2) the Burn-Cut with parallel holes.

1.1.3 TAPER PLUG

In the taper plug **(Fig. 7.3)** inclined holes on the surface of the working face are

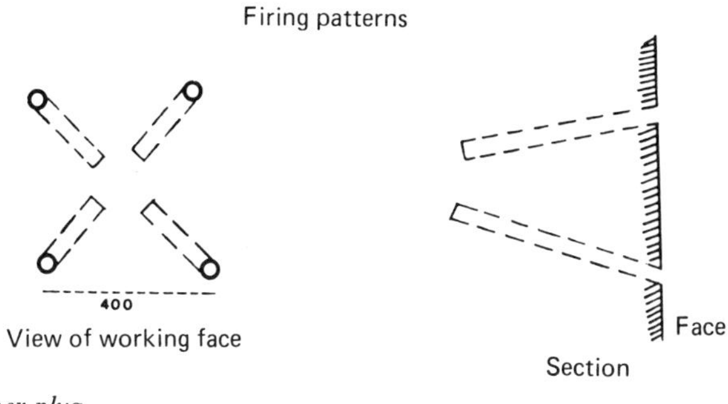

Fig. 7.3. *Taper plug.*

drilled to be convergent and form the outline of a cone or the edges of a pyramid. Removal of the plug demands considerable use of explosive. Holes are then drilled in concentric series. The first series is less convergent than the holes of the taper plug. The second may be formed of holes parallel to the axis of the gallery, and the last holes, the so-called 'scouring' holes, are slightly divergent. This ancient and traditional process is still widely employed. It nevertheless has the drawback, especially in narrow galleries, of not allowing long working lengths because the workers have to drill at an angle to the walls in an inconvenient position, and the dimensions of the hammer limit the depth of the holes that can be drilled.

1.1.4 BURN CUT

This is also called the parallel plug or Canadian plug **(Fig. 7.4).** The process consists of drilling all the holes, parallel to the direction of the gallery, but leaving certain

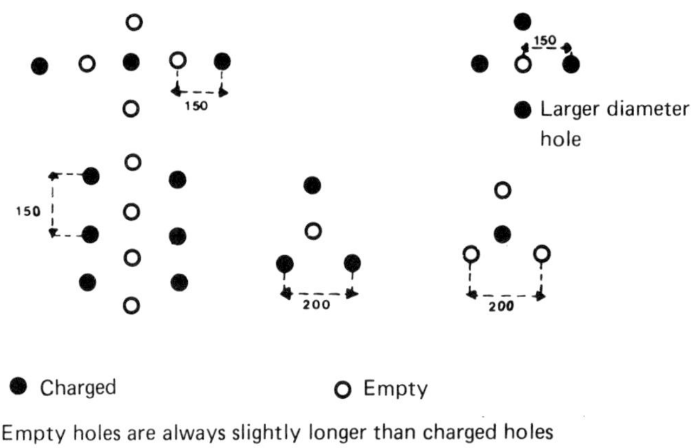

Fig. 7.4. *Burn cut.*

voids between them around the centre. These holes, which contain no explosive charges, play the role of release surfaces by allowing the rocks to expand. Many variants of the burn cut exist. They differ mainly in the layout of the empty holes. These holes are sometimes much larger in diameter than the charged holes, favouring the release of the material.

The use of the parallel plug facilitates drilling, especially with the use of so-called 'jumbo' battery machines. There is no longer any need to find the direction of the holes by trial and error because it is the same for all. Moreover, the hole can be very long (normally up to 3m) and working more extensive. However, this process requires a greater use of explosives.

1.1.5 BLAST FIRING

As described above an overall firing pattern includes the plug. Since the plug has the first blast of the firing pattern, this opening must be widened to the desired cross-section by firing 'scouring' holes **(Fig. 7.5)**.

The number of 'scouring' holes varies with the hardness of the rock. The plug holes are primed with short-delay blasts (millisecond) 25/1000 second, whereas 'scouring' or squaring holes are blasted with a ½ second delay. Instantaneous detonators are also available; these are mentioned for information purposes only because they are used less and less.

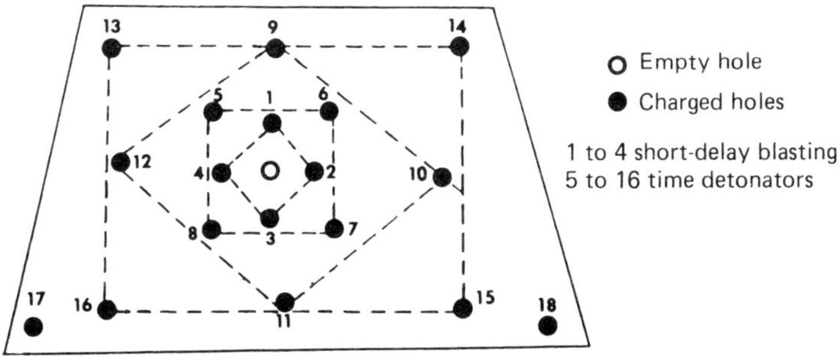

Fig. 7.5. *Firing pattern on working face.*

The advent of time detonators has been a great advance especially in rock driving, by allowing the cutting of the entire gallery cross-section by a single firing blast. Since this avoids returning to the drilling site with the charge and switching of the next volley, very long lengths have thus been achieved. Faced with this advantage the price of the detonators, about 30% more expensive, is a minor drawback.

1.1.5.1 Firing lines

The firing lines consist of insulated wires or, more often, wires placed on insulators. The primers are connected before firing, so as to connect the detonators, generally in series (exceptionally in parallel or in parallel series).

The power source may be a permanent firing network at 100 or 250V (this is the case with instantaneous-release mines in which blasting is controlled from the surface). Nearly always, however, firing is conducted by portable dynamo or battery exploders. Dynamo exploders are the most widely used in France. In these instruments a spring is generally compressed by a removable handle, and the expansion of the spring causes rotation of the dynamo at a specific speed, capable of generating a current in the line greater than the minimum detonation current.

1.1.6 CHARGING OF THE SHOT HOLE

Before introducing the charge the shot hole must be cleansed thoroughly. It is important to prevent rock debris from accumulating between two cartridges, which could stop the propagation of the explosive wave. This is achieved using a metal scraper or a compressed air jet (blower).

The primed cartridge is inserted at the end of the hole (back priming, the only type authorized for time detonators) by pushing it with the help of a wooden or plastic ram. Slight pressure is applied to the ram, but without impact packing. Depending on the depth of the hole additional cartridges are inserted in the same way.

1.1.7 TAMPING

The insertion of the explosive charge is followed by tamping. The tamping normally consists of clay tampings (without gravel) which are introduced in contact with the charge. The first tamping is lightly packed against the charge, and packing is progressive, varying with distance from the charge.

Length of tamping: 20cm for the first 100g of explosive, 10cm for each additional 100g with a maximum of 50cm.

1.1.8 SWITCHING FOR FIRING

All the wires projecting from the blast holes are connected in series to avoid misfires **(Fig. 7.6)**. The electric current flows through a closed curcuit and passes through all the cartridges. In case of a defect the volley is not fired. The resistance of the circuit should be checked before each firing using an ohmmeter.

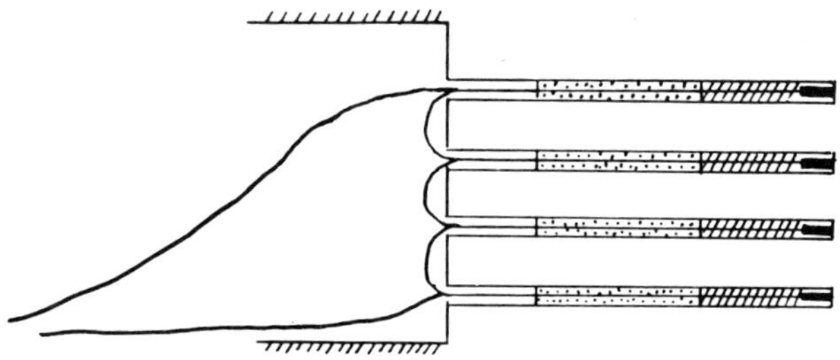

Fig. 7.6. *Firing connections.*

Ohmmeters have recently been replaced by firing circuit continuity testers, allowing a check of the electrical circuits at the working face, in the immediate vicinity of holes charged with explosives and ready for blasting. This tester is a selenium photo-emissive cell whose electromotive force remains so low that it cannot generate any dangerous current, even in a low intensity detonator.

1.1.9 MISFIRES

If a shot hole fails to blast, a second attempt can be made. If the wires are intact, but the attempt fails, the normal method provided by regulations consists of firing a replacement blast, in a second shot hole drilled 20cm from the charged hole. This practice is not entirely risk free:

1) if the hole is too close and deviation takes place, an explosion may occur;
2) if it is too far it will not detonate the explosive of the misfire and unexploded cartridges and the primer will be projected with the debris.

Hence it may appear preferable, as authorized by the regulations with certain reservations, to remove the tamping carefully, and to place a new primed cartridge in contact with the charge. Blasting is then carried out.

1.1.10 BOTTOMS

When a shot hole has failed if the hole is intact along its entire length, it is said to have a 'barrel'; if the hole only subsists on a fraction of its length, this is referred to as bottoms. Holes which have resulted in barrels and bottoms must be identified immediately by means of wooden pegs.

Bottoms can be recharged when they are large, but it is necessary to wait half an hour after the return to the mine site before placing a cartridge in contact with an explosive fragment in slow combustion. However, it is better to avoid re-using the bottoms and to drill holes nearby. At all events it is absolutely prohibited to deepen bottoms and to scour them. Nearly half of all blasting accidents are actually caused by the deepening of unexploded holes.

1.1.11 STORAGE OF EXPLOSIVES

In France it is unauthorized for any individual to buy and use explosives if he does not possess a depot to store them, approved by the Bureau of Mines. Detonators and explosives must be stored in different depots. All the depots must be under the permanent surveillance of a guard. Three types of depot exist in France, depending on the amount to be stored:

1) over 250kg;
2) between 250kg and 50kg and less than 50kg of dynamite. For the first two categories surface depots must be surrounded by an earth barricade; depots in the third category (under 50kg) are easy and cheap to build. For the undertaking of certain short-duration works it is possible to secure authorization to use a mobile or temporary depot.

1.2 Supports

Underground excavations disturb the equilibrium of the ground around the voids created (galleries) and the latter tend to close again, so that the supports are designed to keep these voids open for the time required to complete the works.

1.2.1 GALLERY TIMBERING

The oldest support procedure employs wooden frames. For small galleries like the ones discussed here, which do not need to be kept open for an extended period, timbers 15 to 25cm in diameter are adequate to guarantee preservation during excavations.

Timbering **(Fig. 7.7)** consists of two props and a cap. These components are assembled in two ways:

1) The two ends of the cap and the top of each prop are notched (double notch). This procedure requires the use of a saw. In addition, the props and the cap must be cut to the desired length so that the frame bears tightly against the gallery.

2) The cuts at one end of each prop are easier to cut, and a machete or hatchet is used for the purpose. One starts by cutting this end into a flattened dihedron with a 160° angle. A groove is then cut out of the cap to match the cut end of the prop. Installation is carried out by striking each of the props to force the cap against the top of the gallery. For galleries intended to remain open for a very short period (for

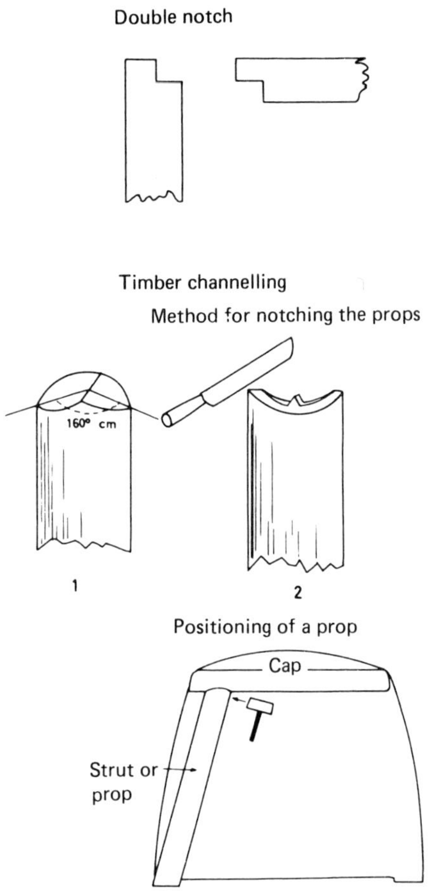

Fig. 7.7a. *Double notch gallery timbering.*

the works only), there is no point in putting a frame between the two props, and a wooden spreader under the cap designed to cope with lateral forces of the sides is sufficient.

The essential requirement for any timbering is that it should be perfectly tight against the top of the gallery directly above the props. This method does not require special qualifications concerning timbering and is therefore the most widely used by the prospector. A lagging of untrimmed boards or laths is placed above the frames to protect the workers against small falling stones.

1.2.2 TIMBERING OF SMALL SHAFTS

We are concerned here only with small exploration pits **(Fig. 7.8)** sunk in overburden shafts or rock, with poor strength, as the sinking of shafts in alluvium has been described in detail in Chapter 3. This chapter is referred to for matters of safety and distribution of cuttings.

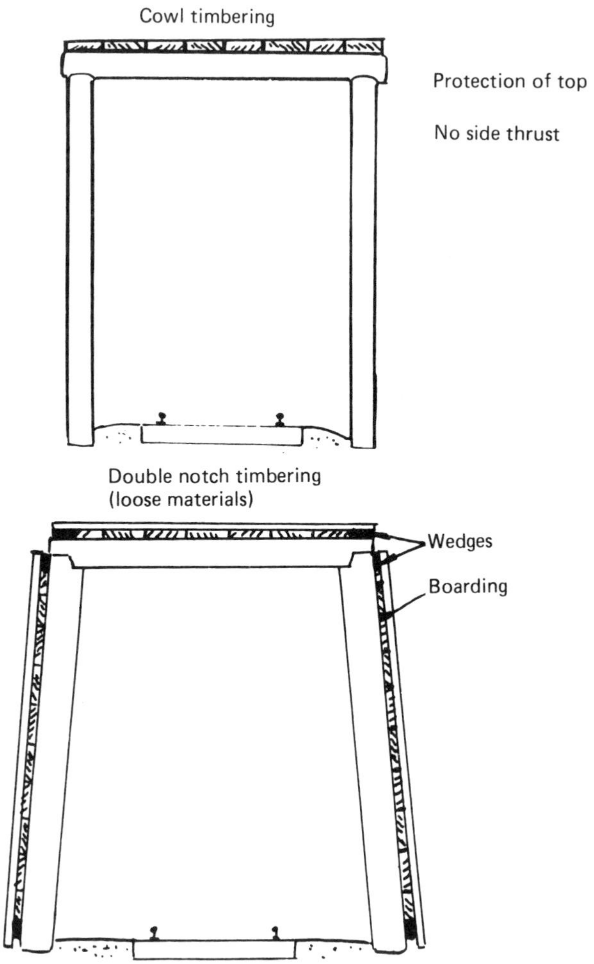

Fig. 7.7b. *Channel gallery timbering.*

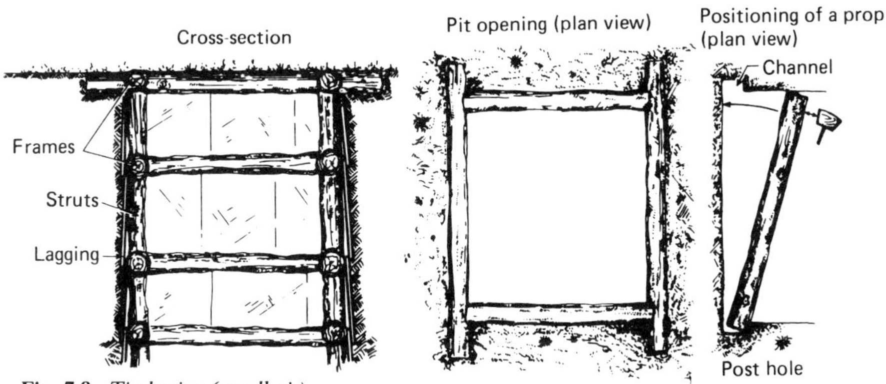

Fig. 7.8. *Timbering (small pit).*

It should first be noted that whenever possible, a pit should be dug without timbering. Always begin by testing the strength of the ground by digging a circular hole 60 to 70cm in diameter. If the ground holds, proceed further, and if not, the pit can be widened to obtain a square cross-section with approximately 1.50m sides.

The timbers at the mouth of the shaft are placed in grooves projecting some 20cm from the sides of the shaft. These timbers are pressed against the wall of the shaft by other notched timbers (see plan view). Inside the shaft a post hole (hole slightly larger in diameter than the frame) about 10cm deep is dug in a wall against the angle with the other wall. At the same level on the opposite wall, a channel equal in width to the diameter of the frame is excavated, about 30cm long and varying in depth from 0 at a distance of some 30cm from the angle to 3 to 5cm against the angle. One end of the timber is placed in the post hole and the other end is pushed into the groove by striking the wood (see sketch). Once both props are in place in the holes at the same level on opposing walls of the shaft, they are pressed against the shaft by notched timbers. This forms a frame which is made integral with the frame immediately above by struts in each corner of the shaft. As with galleries, lagging is placed between the frames and the pit walls.

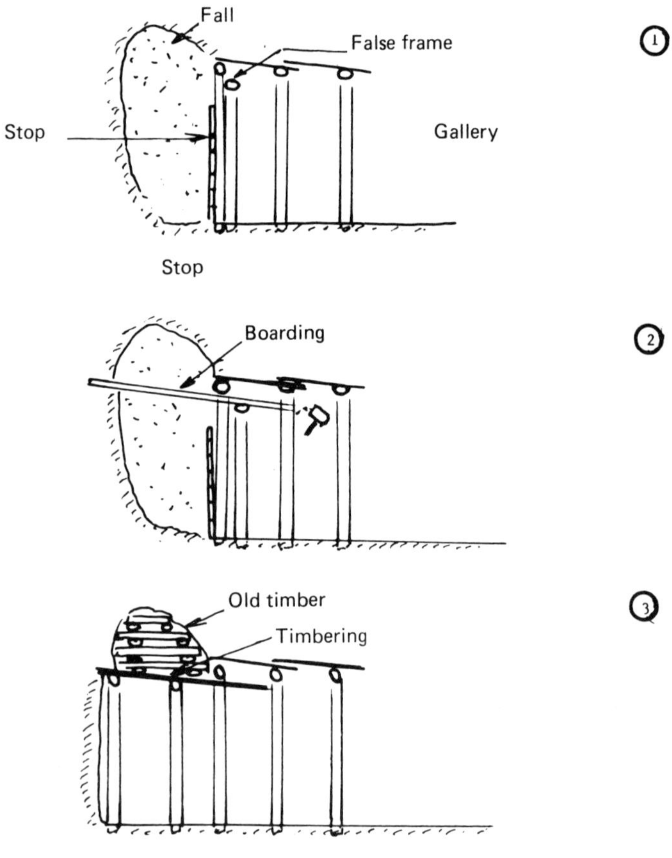

Fig. 7.9. *Passage of a fall.*

The struts prevent the overall timbering from being demolished by the blasts by making all the frames integral with each other. They enable the timbering to follow the sinking of the pit practically up to the working face without undergoing serious damage. However, it is preferable to reduce them to the minimum, by adjusting the blast carefully, namely by using the minimum amount of explosive. Good results are achieved by reducing the length of the blasts on the one hand (0.80m maximum per blast) and by increasing the number of empty holes in the burn cut.

1.2.3 METALLIC SUPPORT

When the ground pressures are high wooden supports tend to break. The ground can be held better with sliding metal supports (Toussaint-Heintzmann type).

1.2.4 TOUSSAINT-HEINTZMANN METAL FRAME

This is the most widely used metal frame **(Fig. 7.10)** and can be installed very rapidly. The TH section offers the special asset of considerable lateral bending strength, while retaining outstanding vertical bending strength.

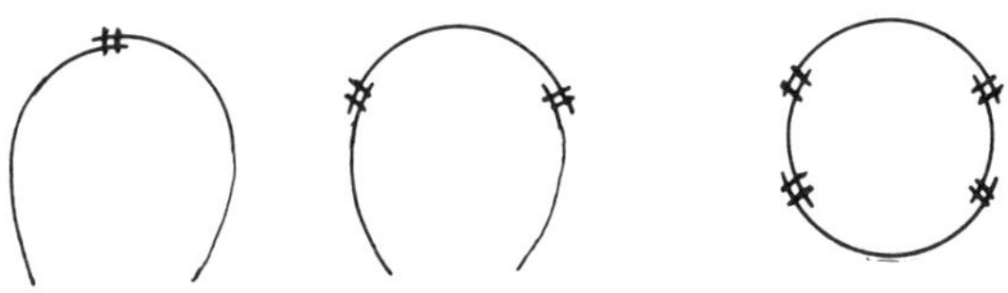

Fig. 7.10a. *Metal support.*

These metal frames consist of two, three or even four elements. These elements slide against each other and are made integral by clamps. The clamp nuts are tightened by a torque wrench. The success of the frame depends not only on the installation, but on two other very important factors, namely the lining and connection of the frames. If the lining is deficient, the frame bends. As for the absence of a lateral link this can cause the arch to slide and the section to leave the plane of the frame. The lining may consist of planks or untrimmed boards and the frames are connected by braces **(Figs 7.10b and 7.11).**

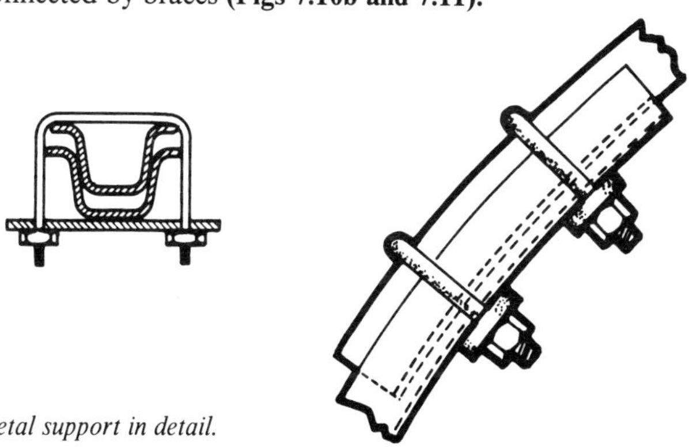

Fig. 7.10b. *Metal support in detail.*

The braces consist of channel sections in which the notches are cut corresponding to the passage of the frames. The brace is clamped to the frame by rounds cut on one side to bear on the frame and threaded on the other for bolt fastening. The braces clamp the frames together and prevent them from collapsing under the projections of blasts.

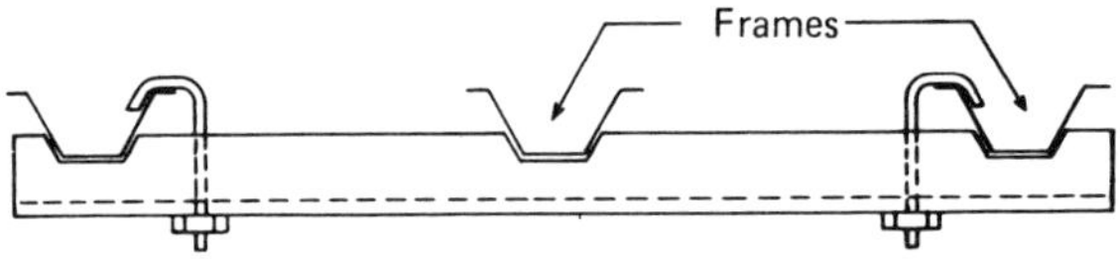

Fig. 7.11. *Detail of lagging.*

1.3 Removal of waste rock

Waste rock mined manually by pick or pick hammer, or by explosive, must be moved to the surface. For small-scale mining works the most widely used method consists of installing a narrow gauge track (0.60m) on the floor of the gallery to convey 100 litre or larger mine cars which, once loaded by shovel at the working face, are pushed by hand to the outside, where they are emptied by dumping. To facilitate the rolling of full mine cars the slope descends slightly from the working face to the surface (3 to 5mm/m). In case of a water inflow at the working face this slope also allows the water to flow out through a gutter excavated on one side of the gallery.

For larger mining works, and especially in countries where labour is expensive, haulage is mechanized. Trackless loaders are more and more widely used for this purpose.

With this equipment the cross-section of the gallery to be excavated for the larger mine sites is about $9m^2$ or slightly more than with the older equipment (track and mine cars). However, this increase in cross-section is not achieved to the detriment of the cost per metre excavated. This method helps to obtain a larger gallery at lower cost, much more easily re-usable in case of subsequent mining. The excavation of these galleries is always entrusted to mine foremen, and the prospector in charge of geological surveys, sampling, etc.

1.4 Gallery ventilation

Ventilation is designed to guarantee a mine atmosphere with a composition, temperature and relative humidity compatible with the safety, health and output of the personnel, and therefore has a three-fold objective:

1) to guarantee that the personnel can breathe;
2) to dilute the harmful gases (firedamp, carbon dioxide, carbon monoxide, nitrogen oxides);
3) to lower the temperature in certain mines.

1.4.1 RESPIRATION

A man needs 25 litres of air per minute for respiration. The air expelled by workers contains 0.4% CO_2. This spent air is remixed and breathed by the workers active further downstream. To ensure that the air does not lose more than 1% oxygen 100 litres of air per worker and per minute are required.

1.4.2 VENTILATION

Adequate and stable ventilation **(Fig. 7.12)** can only be achieved in a mine by the use of fans. Two main types of fan are employed:

1) centrifugal fans;
2) screw fans.

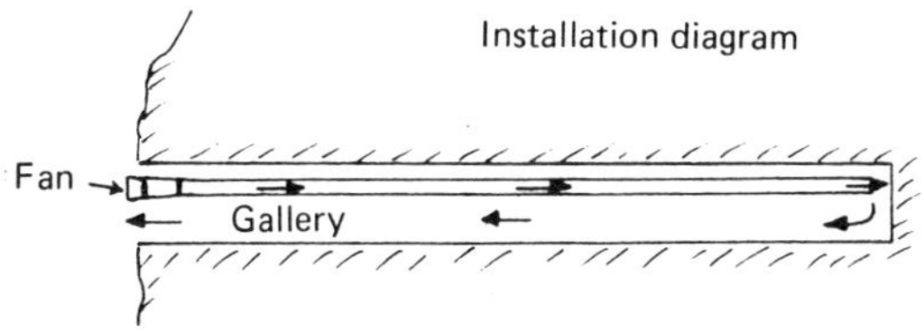

Fig. 7.12. *Installation diagram.*

At present the screw fan is practically the only one used for the ventilation of mines.

In the screw fan the rotor has a horizontal hub to which a number of vanes are fixed. The air enters and leaves parallel to the machine in a helical fashion. The fans are normally driven by electric motors. These fans are installed at the air inlet of the mining works, and the pulsed air flows in a metal or plastic duct (400 to 800mm in diameter) up to the working face, with the gallery serving for the air return.

Fans with a capacity of 5HP can be used for small scale mining works, and can be connected in series. For the excavation of small relatively short exploration galleries, the use of a fan can be supplanted by the installation of blower gas vents **(Fig. 7.13)**. A blower gas vent consists of a 100mm diameter pipe and operates according to the venturi principle.

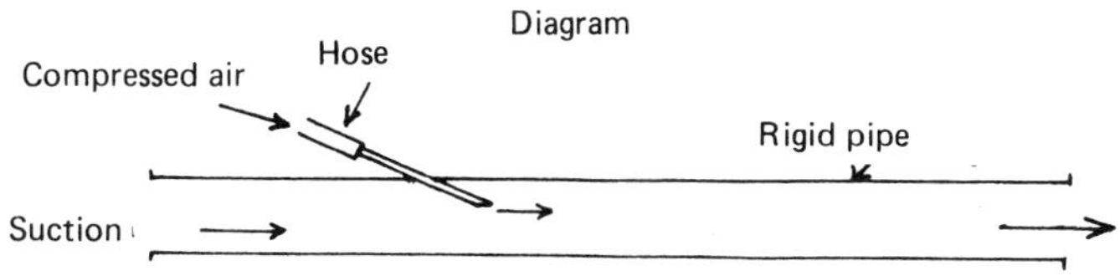

Fig. 7.13. *Blower gas vent.*

For very long galleries this fixed installation, which can operate during blasting operations, allows access to the mine site for the repositioning of the ducts or pipes of the main ventilation system, shortly after firing.

For the ventilation of small shafts it is preferable to avoid the use of explosives in small exploration pits whenever possible. If not, the shaft can be ventilated by lowering compressed air hoses to expel the waste gases.

Even in shafts sunk without explosives, beyond a certain depth (about 10m), it is necessary to expel the gases from the bottom, even if inert, because they can reduce the oxygen content of the air. If compressed air is unavailable, the process used consists of preparing a bed of branches of the same diameter as the shaft (broom is often used in France), introducing it into the shaft at the end of a rope, and removing it to the surface for shaking before it is lowered again. This operation is repeated three to four times before descending into the shaft.

It is recommended that a candle is lowered or burning paper is thrown into the shaft to make sure that the oxygen content is greater than 17%.

1.5 Drainage

Rain water flows at the surface to form rivers. However, part of the water is absorbed by permeable soils before reaching the river beds, forming waters which lie on an impermeable lower bedrock. These waters flow in the ground. They find exits at the low points, thus forming springs. The water-table is not horizontal and tends to assume the form of a curve parallel to the topographical surface. Faults and fractures break the continuity of the impermeable beds and thus lead to the infiltration of water into mine workings.

During exploratory mining works, below the water-table, sudden water inflows are often encountered in contact with faults, veins and so on. To avoid inundating the works under way by a sudden rise in the water levels suitable means must be provided to remove the water to the surface and to allow the work to proceed normally. Pumps are used for this purpose. For mineral exploration the pumps are of two types:

1) compressed air;
2) electric.

1.5.1 COMPRESSED AIR PUMPS

In all cases, pumps must be used for muddy waters. Turbine pumps driven by a vane motor are the most widely employed, actuated by compressed air. Through the intermediary of a rotor fitted with vanes, the compressed air drives a turbine (pumps: Renault- Rateau, Ingersoll, Atlas-Copco). A semi-rigid or flexible pipe connected to the discharge end transports the water to the surface.

In exploration operations the pumps should have a short discharge head but with a delivery exceeding the water inflow. If the pump is not powerful enough to discharge to the desired height, two pumps can be connected in series, and the delivery is thus maintained at the head multiplied by two.

1.5.2 ELECTRIC SUBMERSIBLE PUMPS

These vertical pumps **(Fig. 7.14)** consist of an electric motor driving a turbine, some of which are equipped with a special cooling system, and are capable of operating with no load (Flygt). They are calculated as for compressed air pumps, and can be connected in series to increase the discharge head.

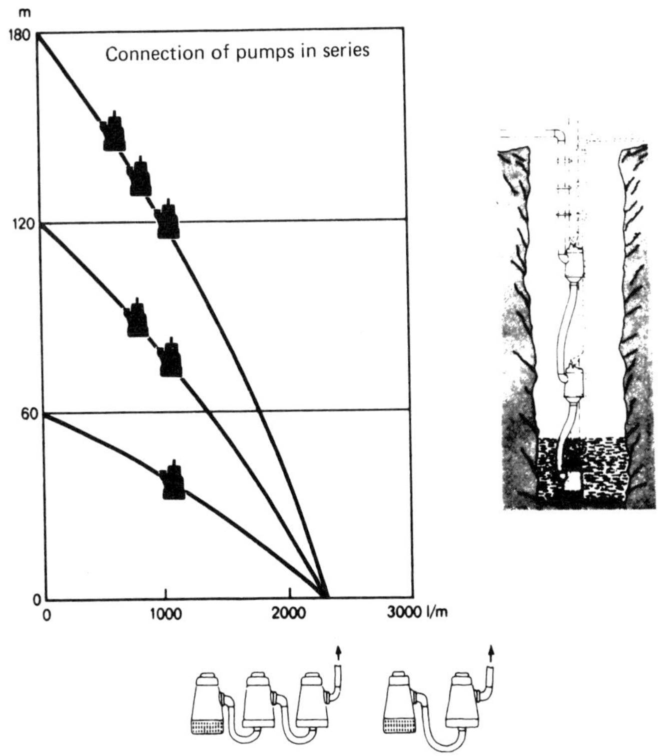

Fig. 7.14. *Connection of submersible pumps.*

1.6 Lighting

Lighting raises a particular problem in mines because of:

1) transfer of the work sites;
2) working conditions;
3) minimum reflecting power of the gallery walls.

Good lighting improves:

1) safety;
2) output;
3) product quality.

1.6.1 LIGHTING DEVELOPMENTS

1.6.1.1 Acetylene lamp

The acetylene lamp is not a safety lamp. It is light-weight, practical, offers high light intensity at low cost, but its operation is irregular, and it can start fires due to its naked flame. This lamp, hitherto extremely widespread, is now being discarded in favour of the electric helmet lamp.

1.6.1.2 Helmet lamp
The battery, placed in the metal or plastic case, is carried on the belt, and connected by a flexible cable to a small bakelite projector attached to the helmet, which contains a main bulb and an auxiliary bulb. The lamp must be recharged after each work shift.

Two types of battery are available, acid batteries and alkaline batteries. Both alternatives are comparable from weight, safety and lighting standpoints.

1.7 Heavy equipment

Most mining work sites require two energy sources for equipment operation:

1) compressed air (compressors) for working, drilling and drainage equipment;
2) electricity: motor generator sets, ventilation, drainage, lighting.

Three main types of compressor are available:

1) piston compressors (the oldest);
2) rotary compressors (Ingersoll, Worthington);
3) screw compressors.

1.7.1 CHOICE OF A COMPRESSOR

1) The capacity of the compressors must be calculated in accordance with the required air flow rate.
2) The means of access to the work site must be taken into account. If access is difficult, light-weight compressors must be provided, with a larger quantity if necessary to obtain the required flow. If two or more compressors are required, to ensure their correct operation a compressed air tank must be provided to avoid jerky operation and to standardize the flow **(Fig. 7.15)**.

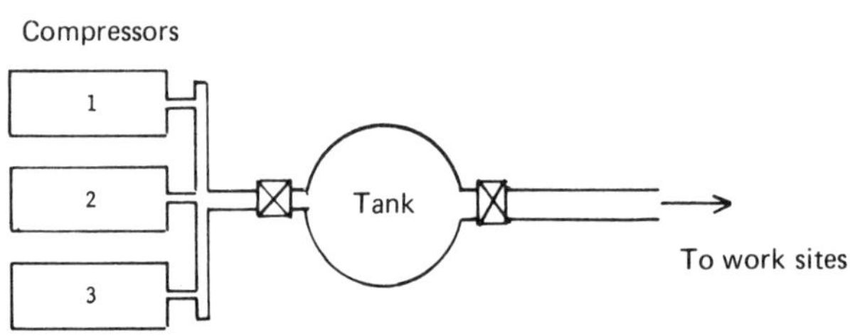

Fig. 7.15. *Compressor installation diagram.*

Motor generator sets are used at large mining sites and camps; their selection depends on the needs of the site and is beyond the competence of the prospector.

2 Sampling in mining works

Whether for the blasting of outcrops, shafts and prospecting trenches (Chapter 2), in pits or drill holes in detrital deposits (Chapter 3), in percussion or cored bore holes for deposits in place (Chapter 6) or for mining works, sampling is the prospector's most important work. It must be performed with the greatest care from the taking of the samples to their reduction to a sufficient quantity for analysis. Obviously the latter must also be extremely thorough, but lies outside the prospector's field of activity. The decisions concerning further operations to the survey of showings or deposits will be sound or unsound depending on the representativity of the sample.

2.1 Survey of galleries, advancement log, sampling plan

In a gallery or shaft being excavated the geological and possibly topographical survey is conducted at the same time as sampling, each working face, characterized by its position in relation to the entrance of the gallery and by the date on which it is reached, is the subject of a sketch including the main geological and mineralogical observations. If it is channelled for sampling, the position of the channel and the sampled lengths are recorded. If not, the characteristic dimensions are measured such as thickness and reduced thickness. All this information is transferred to an advancement log which is subdivided according to the number of galleries in progress. To each subdivision is appended a drawing of the gallery concerned, showing its origin and the direction of geographical or magnetic North, with a note of the date of the topographical survey.

The results of analyses of samples taken are transferred to their place in the log and on the appended drawings.

For older mining works the geological survey of the gallery is conducted after having washed the walls (or one of them) and the ceiling with water. If a large-scale gallery drawing exists (1:100 to 1:500), the survey results are transferred there too. If not, a rapid topographical survey of the gallery is conducted with a compass and 'topofil'.

The data collected serve to determine the sampling plan of the gallery, if any. When the analysis results are obtained they are transferred to the drawings **(Fig. 7.16)**.

2.2 Sampling

Sampling operations in mine workings can be performed in three ways:

1) on piles mined as the gallery is driven;
2) on drilling muds of shot holes,
3) by channelling.

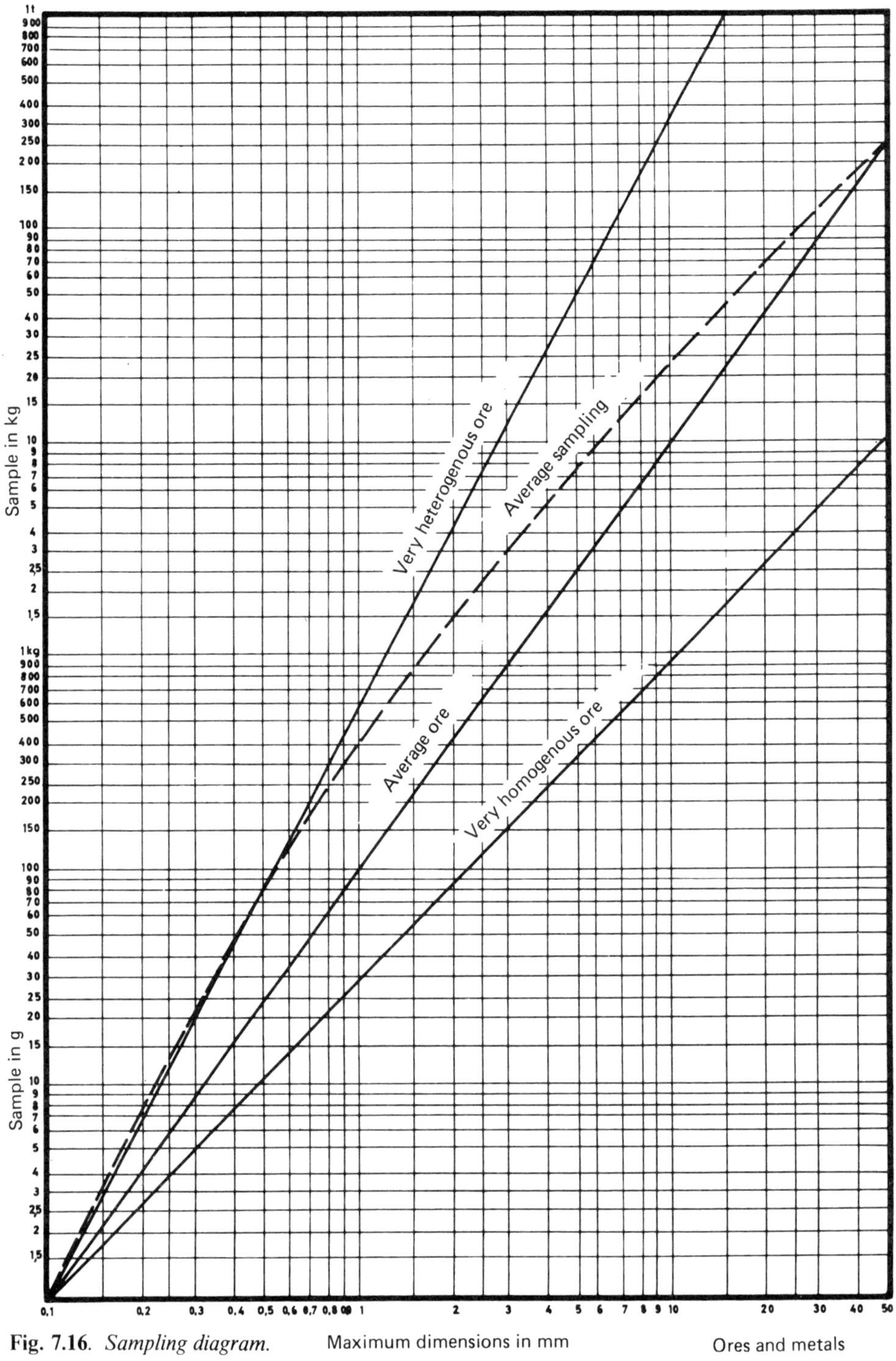

Fig. 7.16. *Sampling diagram.*

Before examining them in detail some theoretical indications are in order on the distribution of valuable minerals in the gangue.

The ores may be homogeneous or heterogeneous. An ore consisting of a single useful substance without gangue is a perfectly homogeneous ore. An ore in which the useful substance is only concentrated at a few points in a large volume is an extremely heterogeneous ore. A whole range of ore homogeneities exists between these two extreme cases. Among homogeneous ores are the heavy ores (iron, manganese, etc.). Among heterogeneous ores are those of the precious metals (gold, silver, platinum), radioactive metals and relatively rare metals (tungsten, molybdenum, niobium, tantalum, etc.) and also generally pegmatites. The ores of base metals (copper, lead, zinc etc.) are generally average ores.

It is clear that the more homogeneous an ore the greater the chances of the different pieces making it up being similar, even if their grain size is coarse. This

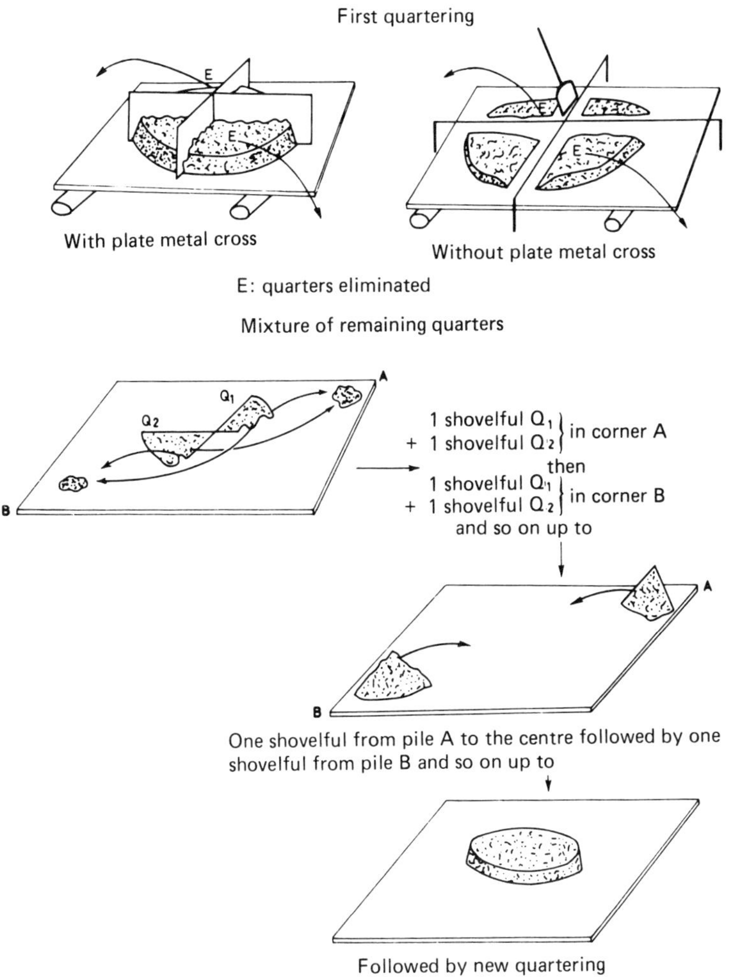

Fig. 7.17. *Initial reduction operations (trench sample).*

eliminates the need to grind it as fine at equivalent weight as a heterogeneous ore. Likewise at equivalent particle size distribution, a smaller sample is needed. This clearly emerges from the sampling diagram **(Fig. 7.17)**. This diagram is only approximate and must be replaced for greater accuracy by the sampling calculator (normogram) of P Gy, which is more complicated to use. As it stands it nevertheless helps to avoid blatant errors in assessing the fineness of the sample grains to be reduced.

2.2.1 SAMPLING FROM MINED PILES

Two cases are examined:

1) sampling of measured piles of cuttings from prospected pits;
2) sampling of earth from a firing blast at the working face of a gallery.

2.2.1.1 Pit spoil

As described in Chapter 2 the material taken from each metre advanced in a pit are placed in a spiral around the pit. These pits are often sunk on showings or deposits of heavy ores (bauxite, iron, manganese) which are generally homogeneous. Each measured pile of an 80cm diameter pit represents 500 litres of a sample in place corresponding to weights ranging from 1000kg for bauxite to 2000kg for iron or manganese oxides. Each measured pile is reduced separately.

One begins by cleaning an area with a radius of about 1m, its centre located 2m from the pile to be reduced. If this is too difficult or time consuming, a thick plate is positioned at this location, with an area of 3 to 4m^2.

Two labourers stand on each side of the original pile, and extract shovelfuls which they throw alternately on the cleaned area or plate. Large blocks are temporarily set aside for reduction by sledgehammer to dimensions of the size of a fist (about 10cm) and subsequent mixing with the pile formed at the end of shovelling.

The new pile is flattened in the form of a circular disc and quartering operations are performed as shown in Fig. 7.17 until a sample of about 20 litres is obtained. If the ore is an average ore, the following operations **(see Fig. 7.18)** may be necessary for an initial weight of 1400kg:

1) quarterings on the ore at 10cm, new weight about 700kg;
2) crushing to 5cm;
3) two quarterings, new weight about 180kg;
4) crushing to 2cm;
5) two quarterings, final weight about 45kg.

The sample thus obtained is sent for sample finishing with a label bearing the sample number, sector name, pit number, place of sampling in the pit, date and author. One half of the sample is sent for analysis and the second serves as a reference.

2.2.1.2 Materials from blasting in a gallery

The sampling of materials mined during the blasting of a gallery takes place on

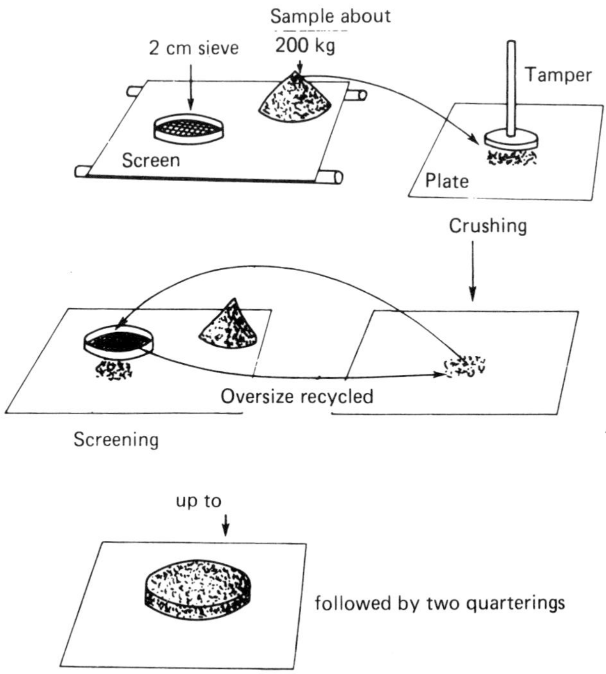

Fig. 7.18. *Sample reduction (average ore trench sample).*

loading, either by taking one shovelful every n shovels loaded, or by taking one mine car from n mine cars extracted.

In the first case, the shovelful selected is set aside in a half-drum. This process has the drawback of delaying loading. The number n may vary between 3 for a heterogeneous ore and 50 for a homogeneous ore. For a medium ore it is about 10. For a gallery cross-section of $4m^2$ the weight of materials should be about 10 to 15t (or more depending on density)/m of advance giving a sample of 1000 to 1500kg for an average ore, or 5 to 8 half-drums. Each of these half-drums receives the same label indicating the sample number.

In the latter case, loading takes place normally, and the mine cars are transferred to the surface. In principle, the number n of mine cars is the same as the previous number n of shovelfuls. In actual fact, the number of mine cars from one blasting being only by chance a multiple of 10 (for an average ore for example), n is generally taken as less than (or equal to) 10. The selected mine cars are emptied on the area reserved for the rest of the sampling operations. This sampling method in no way hinders the clearance of the working face. However, it is liable to yield erroneous results because the materials from one blast are distributed on the ground in accordance with prevailing conditions, and are not mixed. An experiment was conducted on a complete blast in which 16 mine cars were loaded and analysed systematically, in accordance with all the cleanliness and sample reduction standards. For an average content of 0.459% Pb one mine car gave 1.25% Pb and another 0.25% Pb, or two to three times the average, whereas for an average content of 5.028% Zn the same mine cars gave 9.40% Zn and 3.25% Zn, differences as wide as plus 87% and minus 35% in relation to the mean.

By systematically taking one mine car out of 8 as the blast sample (the first and ninth, followed by the second and tenth, etc.) the differences to the mean content of the 16 mine cars are still high or a maximum of plus 63% and minus 24% for Pb, and plus 27% and minus 34% for zinc. With one mine car out of five, this difference is redued to plus 22% and minus 5% for Ph and plus 11% and minus 1% for Zn. For Pb it is probable that the accuracy on the analyses exerts a greater effect on the relative error given the low contents prevailing.

At all events, to avoid an excessively high systematic error, assuming that one mine car is sampled out of four, for example, in the experimental case, mine cars 1, 5, 9 and 13 are taken for the first group, 2, 6, 10 and 14 for the second, 3, 7, 1 and 15 for the third and 4, 8, 12 and 16 for the fourth group, and the cycle is then repeated.

Sampling from mined piles has the drawback, especially for average ores, of requiring operations on large quantities of ore and proceeding with intermediate crushings to respect the particle size distributions indicated in P Gys monograph, or in Fig. 7.16. These drawbacks are significantly reduced for homogeneous ores, especially heavy ores, for which sampling from mined piles appears to be satisfactory. The samples are prepared in the same way as the spoil from a pit.

2.2.2 SAMPLING ON DRILL HOLES

In the course of driving galleries it is preferable, for average ores (base metal ores generally) to take samples from the muds leaving the blast holes drilled in the working face. This method, in use for many years, but adapted by the BRGM to rapid exploration sites, offers the two-fold advantage of disturbing the blasting of the gallery to a minimum and supplying relatively light samples with fine particle size distribution, thus avoiding the use of crushers. All the cuttings that can be collected from the drill holes (25 to 40) of one 1.50m blast represent (for a cross-section of 3.5 to 5m^2) about 1% of the total weight of the materials blasted, namely 150 to 200kg.

This method is ideal for scattered coarse mineralization distributed over the entire working face.

Prior to any operation the recovery funnel **(Fig. 7.19)** is fastened to the base of the working face (0.5m from the floor) using two iron bars inserted in two holes drilled for the purpose. The irregularities of the gallery face at the funnel level are filled by boards and stones. Working face/funnel back connection and tightness are achieved by the use of plaster. The width of the funnel is calculated so that all the run-off on the face is collected by the funnel. Anchoring by means of plaster is soon solid enough to proceed with thorough washing of the working face with a water jet, and also of the equipment already in place. The recovery tank is then placed under the outlet of the funnel. Shot holes can then be drilled as if no sampling was taking place, and it is simply necessary to remove the tank and the funnel when shot holes are drilled in the floor.

It is obviously unnecessary to sample all the holes drilled to obtain a good distribution of samplings. Since the firing patterns are extremely variable, more detailed information cannot be provided, but it may be remarked that very often,

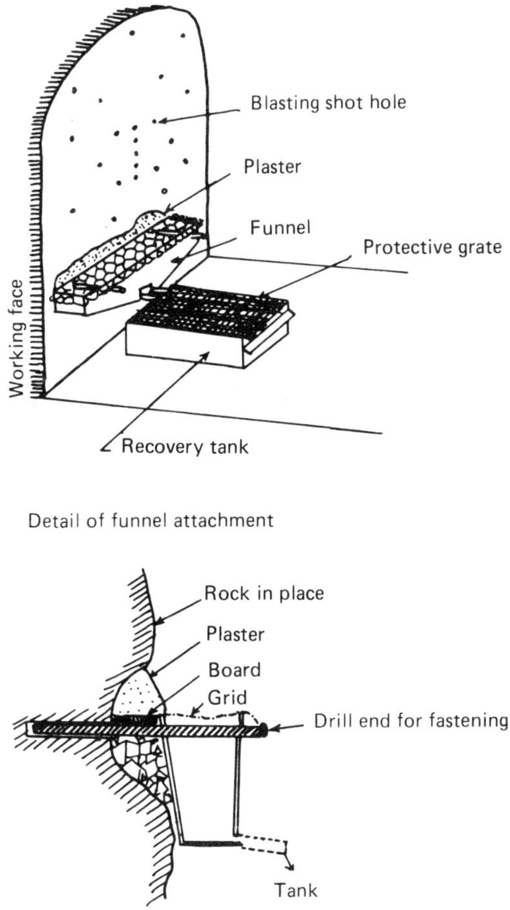

Fig. 7.19. *Sampling of drill holes.*

it is advantageous to sample one hole out of two. The failure to sample a hole corresponds either to removing the tank during its drilling or to performing this operation when all the holes to be sampled are drilled and the sampling equipment removed from the face.

A very important prerequisite is that the rock is solid and does not disintegrate under the action of the drilling waters trickling on the working face. Any disintegration is liable to result in the settling tank receiving materials that distort the sampling, especially if the face is composed of hard parts (quartz, limestone, etc.) connected with other soft and friable materials (clay, ore, etc.). In this case, the collector method developed by the BRGM **(Fig. 7.20)** can be employed.

Recovery by collector can also be employed on special drill holes in the top of the gallery but care must always be taken to secure the collector in the preliminary hole by means of wooden wedges.

Hence on this type of scattered mineralization on the entire face, without a

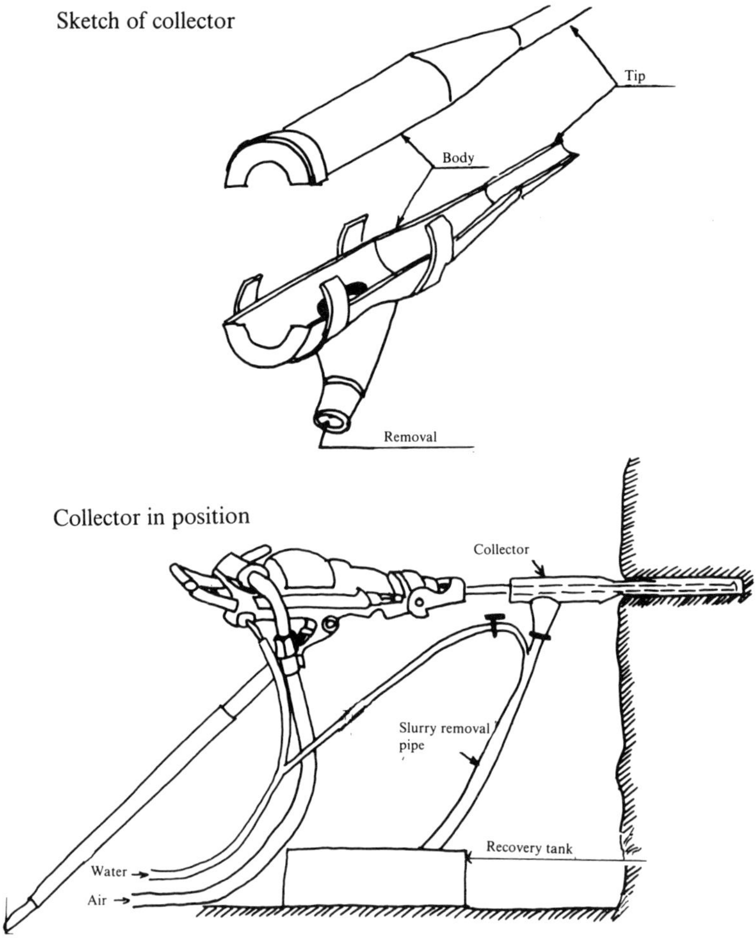

Fig. 7.20. *The BRGM collector mounted on hammer drill.*

pronounced directional character, the drilling muds are sampled from all the shot holes themselves, except the floor holes below the funnel. However, if mineralization occurs differently, special holes may have to be drilled.

If mineralization only occupies part of the face, the following is necessary:

1) if it occupies more than half, operate as above but only collect drilling muds corresponding to the mineralized part;
2) if it occupies only a limited part (thin mineralization for example), proceed with four special sampling holes drilled sideways from one of the sides of the mineralization, at intervals of 25cm along the vertical. Each hole cuts into mineralization, slightly upward, and its length depends on the thickness of mineralization observed on the face (the direction of the hole and its inclination are given by a special table). These indications are

recorded by the individual responsible for sampling, as well as the length of the sampling hole. In many cases, formations of different types and mineralization itself yield muds of different colour, making it possible to determine very accurately the moment when the mineralization zone ends.

If the mineralization is thin and with clearly defined surrounding beds, sampling can also be conducted by channelling. If mineralization is in strips or directional, the following may be necessary:

1) if it is thin, proceed as above by cutting special shot holes;
2) if it occupies all or nearly all of the working face, proceed with special divergent shot holes drilled from the centre of the gallery according to the same principle as above. The sampling may be limited to the width of the gallery, or may spread beyond to sample the entire formation if the latter is thicker than the gallery.

At all events the muds are collected, settled, dried in an oil-fired or infra-red oven depending on the metal involved, quartered in the sampler (Chapter 2) and sent to the laboratory for grinding and analysis.

2.2.3 CHANNEL SAMPLING

This type of sampling **(Fig. 7.21)** can be performed on a working face, but only in the very specific case of a thin vein, for example. However, its main field of

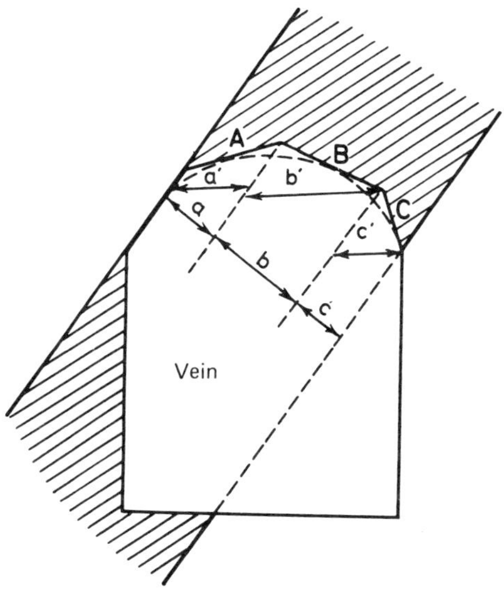

Fig. 7.21. *Subdivision of the channel on a vein in a gallery.*

application concerns suspended mine workings, not equipped with drilling facilities.

The channel sampling (or cutting) technique was described in Chapter 2 in relation to the sampling of a vein in a trench. However, in disused galleries, where a weathered crust may have formed one to several centimetres thick, simple brushing would be inadequate. One needs to cut through the surface to the sound rock and to prepare a relatively flat surface in which the channel can be cut.

In practice, the channels are never cut in the floor of the gallery because it is very difficult to clear a clean area not polluted by ore from elsewhere. This leaves the roof, the two side walls and the working face.

Channels, in the roof for example on a vertical vein followed by a gallery, are difficult to make because of the awkward position of the worker, and require the wearing of special goggles. The debris is collected in a container held at arms length under the channel by another labourer, or on a tarpaulin previously spread on the ground. Precautions must be observed if the gallery is arched because in this case the channel does not cut through all the mineralized strips at a constant angle. The best means is to subdivide the samples as shown in Fig. 7.21, and the contents t_A, t_B, t_C, obtained in channels A, B and C are then weighted by the horizontal distances a', b' and c', or thicknesses a, b and c to give the real content of the vein such as:

$$t = \frac{a' \, t_A + b' \, t_B + c' \, t_C}{a' + b' + c'}$$

In the case of a vein, in addition to the channels, a cut is made in the working face, perpendicular to the vein walls or horizontal, and more rarely vertical for low-dip veins.

When massive mineralization is involved the channels are cut in the side walls. They are either horizontal (also the case for drifts perpendicular to very thick veins) or vertical on the same side at intervals ranging from 1 to 10m depending on the type of mineralization. Sometimesif the two sides display clearly different mineralizations, the channels are placed alternately in each of the sides in order to compare their mean content.

The sample obtained in chips is prepared as described in Chapter 2. For a very heterogeneous ore the approximate grain size to be obtained for a final 100g sample is 0.5mm. In the absence of grinding the total initial sample of about 20 litres to 1mm or less, intermediate grindings can be carried out. For example, for 20 litres (about 40kg) of average ore the first crushing reduces the sample to 5mm, followed by four quarterings which give a sample of 2.5kg, which is then itself crushed to 1mm and then quartered four times to obtain a sample of about 150g. To determine the weight corresponding to these different particle sizes, the sampling diagram is employed **(Fig. 7.16)**.

It should be noted that the same rules apply to samples from the spoil of a pit and blasted materials from a gallery after the crushings and quarterings decribed above, at the end of which the sample weighs between 40 and 50kg.

2.2.4 ESTIMATION OF CONTENTS BY THE MEASUREMENT OF REDUCED THICKNESS

It is sometimes the case that the useful mineralization forms more or less regular strips, veins and lenses in the seam. The term 'reduced thickness' is applied to the thickness that the vein would have were it reduced to the valuable minerals. As with a channel in which only a thin section of the vein is cut out, the reduced thickness on the working face is only measured on a line drawn by chalk perpendicular to the surrounding beds. Along this line, the thickness of each of the valuable minerals is measured and separate totals compiled. Each of these totals represents the reduced

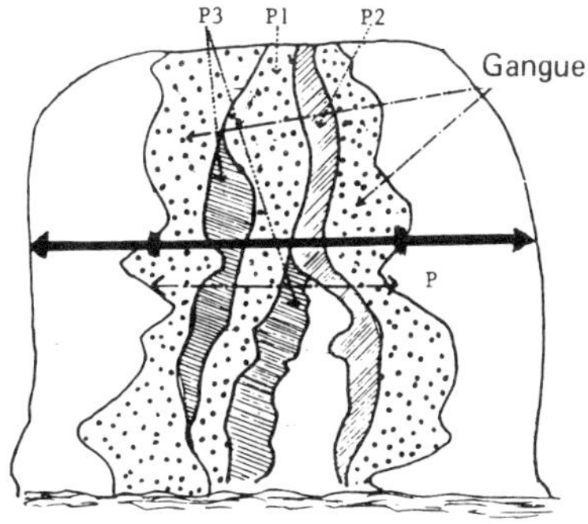

Fig. 7.22. *Estimation of reduced thickness on a working face.*

thickness of the vein in the mineral considered. For a single ore mineral occurring in strips, this operation is easy and the measurement fairly accurate. The operation is more difficult for two minerals, and may lead to dangerous errors above this number.

The contents are calculated from the reduced thicknesses as follows:

let g be the density of the gangue;
d be the density of the pure ore;
both values taken as equal to theoretical values,
P is the thickness of the vein.
p is the reduced thickness of the vein in the mineral considered.
Tm is the metal content of the mineral considered. As a rule, the theoretical value of Tm is taken, which equals

$$\frac{\text{weight of atoms of metal contained in the mineral}}{\text{molecular weight of the mineral}}$$

Here are rounded off values of Tm for the following minerals:

galena	86% Pb
blende	67% Zn
chalcopyrite	35% Cu
chalcocite	80% Cu
stibnite	71% Sb

The weight of ore per unit area is $p \times d$

The weight of gangue per unit area is $(P - p)g$

The mineral content is hence $\dfrac{p \times d}{(P-p)\,g + p \times d}$

and the metal content

$$\frac{t}{100} = \frac{p \times d}{(P-p)\,g + p \times d} \times \frac{Tm}{100}$$

This estimate, which is designed to economize channelling which is always costly, is not usually carried out in parts of the vein where the mineralization is weakest. Even in this case, however, it is important to avoid blatant errors, and every prospector, especially a beginner, should practise the measurement of reduced thicknesses and the estimation of contents by this means, and then cut a channel at the same location to compare his estimate with the results of the analyses.

2.2.5 SAMPLING LOG

A sampling log like the one described in Chapter 2 should be kept for each sample taken in mine workings.

Appendix A1

Technical report

1 General

GEOGRAPHIC DATA

1) For a mineral occurrence: coordinates, access routes, location sketch map.
2) For an area to be prospected: boundaries of the area, surface area, location diagram, etc., access roads, climatic data (best, possible, impossible seasons for work).

LEGAL DATA

1) Holders of rights, partners in the operation, etc.
2) If a proposal has been made by a third party: conditions of the option, and any companies to which proposals have also been made, etc., reasons for refusal, etc.
3) Data of legal, financial, technical, etc. nature of the third party proposing the operation.

CASE HISTORY

1) Previous work by other organizations: known results, tonnage extracted, grades, etc.

2) Previous work by the prospector's own organization: detail by year, work carried out, results obtained, costs.

GEOLOGICAL DATA

1) Regional sketch map.
2) Local geology in the immediate area of the occurrence or anomaly, etc.
3) Description of the occurrence itself, with particular attention to the geometry of the mineralized zones, visible or proven at depth: thickness, length, depth, etc., observed paragenesis, etc. Give the extent and the intensity of the geophysical and geochemical anomalies.
4) Known deposits and occurrences in the area: economic importance, type of deposit, etc.

ECONOMIC DATA

Tonnage (proven, probable, possible), grades, possible by-products. Indicate the reliability of the information according to its source.

BIBLIOGRAPHY

This should be as complete as possible and should include any internal reports. Indicate if photographic cover is available.

ENCLOSURES

Photocopies or summaries of reports and all documents which could provide information for the project.

2 Personal comments and assessment

1) Give details here of all the aspects which have not yet been dealt with; compare the importance of the project with that of others proposed or already carried out, etc.
2) Objective(s) (types of deposit).
3) Summarize the factors (positive, negative or uncertain) which have led to the definition of project objective(s).

3 Proposed work

OBJECT

1) No occurrence or anomaly is yet known.
   - Provisional deposit model(s) (principally bibliographical).
   - Field survey in geological context.
   - General prospecting: search for showing; strategic prospecting: alluvial, geochemical, geophysical, airborne.

2) Occurrences or anomalies are known.
   - First phase: (no significant work carrried out) directed towards deciding whether or not to investigate the occurrence or area of occurrences further. It is important to indicate the kind of occurrence in order to choose the methods to be used later.
   - Second phase (already oriented towards economic development of the occurrence includes): more important work such as tactical geochemistry, ground geophysics and possibly drilling.
3) A deposit exists that needs to be evaluated.
   - Essentially oriented towards specifying ore tonnage and content.

THE PROPOSED OBJECTIVE

This should be limited to the result expected from the phase considered below, and not to the final objective, which is obviously the discovery of an ore deposit.

WORK NECESSARY TO REACH THIS OBJECTIVE

Give enough details to be able to quantify the work, eg number of samples to be taken and their spacing, elements to be assayed for, number and depth of drill holes. Make contact with technical departments for the choice of methods, grids, results to be expected from the methods used, etc.

RESOURCES

1) Physical, personnel, equipment, etc.
2) Services requested of the technical departments.
3) Funds, estimated cost of the work envisaged above.

WORK SCHEDULE

Take into account climatic and other factors.

ANTICIPATED DATE FOR THE COMPLETION OF THE APPRAISAL OF RESULTS OBTAINED FROM THIS PHASE

Appendix A2

Vehicles in desert zones

This document was kindly made available by the security department of the Companie Generale de Geophysique (CGG).

1 *Marking out the tracks*

Besides being necessary from a logistics point of view, clearly marked tracks is the principal precaution against straying from routes regularly taken by expedition vehicles including:

a) access from a public road to a main camp;
b) access from a water supply point to different camps;
c) access from the main camp to secondary camps;
d) access from secondary camps to one or several points on work lines;
e) where appropriate, roads connecting one line to another.

To be effective beacons should:

a) Be close enough to be seen from any other in average visibility for the area (eg 2 to 300m).
b) Be identifiable by a colour or letter code and marked with distances, in kilometres, from appropriate points, say every 5 to 10km.
c) Be made from material which is suited to local conditions: cairns (piles of stones), barrels, plastic tubes, plastic tape, empty bottles, food tins, scrap metal, etc.

To be efficient route-marking should follow field reconnaissance using maps or aerial photographs, so that the track follows the easiest route for all types of vehicle using it.

VERY IMPORTANT

It is imperative that all personnel stick to marked tracks. Whatever the weather conditions, no change of route should be made without the agreement of the head of the expedition and without preliminary reconnaissance before changing the beacons. This procedure is imperative because otherwise the number of tracks and paths would multiply in all directions, causing confusion and complicating any kind of search or rescue operation, even after simple mechanical breakdown.

Steering by windscreen compass should be forbidden except in vehicles carrying a technician working on reconnaissance or opening a track or work line. Putting well compensated compasses in the hands of inexperienced people would make things worse.

Obviously, if work is being carried out in areas (eg stony desert) where the paths remain well defined, beacons may be more spaced out, but the rule of always following the same route still remains to avoid creating tracks in all directions, which can lead to considerable confusion.

2 *Equipment for vehicles*

Vehicles to be used in desert areas must at all times carry special equipment over and above normal requirements:

a) extra spare tyre (except for special vehicles travelling in a convoy);
b) track plates or some other system for freeing vehicles from sand;
c) shovel, tow line, simple repair outfit (fanbelt, hoses, etc.);
d) spare jerrycan of fuel, 5 litres/1 gallon of oil, jerrycan of water if the vehicle is water-cooled;
e) sufficient water for the number of passengers and the travelling time, but with a minimum of 10 litres/2 gallons per passenger, if possible in a tank built into the vehicle;
f) first aid chest;
g) some means of making signals: mirrors, smoke canisters, rockets, etc;
h) the instruction leaflet in the cab.

If possible one of the vehicles in the convoy should have a radio; radio contact should be made on departure, then every three hours during the journey.

As the majority of commercial antifreezes are toxic, it is advisable to empty and rinse out all the vehicles' cooling systems as soon as there is no more risk of freezing. A cooling system must never be used as a possible water reserve because it may still be highly toxic due to descaling products, radiator copper, etc.

It should be noted that each passenger should be equipped with survival clothes even for a short journey and should always be prepared for the possibility of having to bivouac.

3 *Moving from one place to another by vehicle*

FOR SHORT JOURNEYS RETURNING TO THE CAMP THE SAME DAY

A single vehicle will be able to make a journey of more than 25km around a camp or work site only on a beaconed track or line. If the camp, line or a landmark is still in sight, a vehicle may go beyond these limits. If several vehicles leave on a journey, they must do so in convoy and follow the beacons closely.

FOR LONG JOURNEYS ON BEACONED ROUTES

A vehicle may only leave provided that there is one other person besides the driver, that three days spare supply of provisions and water are taken on the journey and that the probable day and time of arrival are known at the destination.

FOR RECONNAISSANCE TRIPS LASTING SEVERAL DAYS

The vehicles may only travel in groups of at least two and someone who can navigate with a compass must be present.

The vehicles must be equipped with a radio and contact the nearest station ahead twice a day, giving their position as accurately as possible.

The route of any trip straying from beaconed roads must be plotted continuously throughout the journey.

In addition to spare provisions, the vehicles will also hold a supply of water and supplies to last 150% of the estimated duration of the journey.

VERY IMPORTANT

It is essential to be very aware of the risk of getting lost on 'routine' trips (to and from the field), which should be approached each time as though it were the first. Trying to open up a new track every day to shorten a journey may lead to getting lost or breaking down in a place which is difficult to find, turning a minor accident into a far more serious incident.

POOR VISIBILITY

In all cases there is good reason to stop if, due to atmospheric conditions, visibility becomes so bad that the beacons are no longer visible, and to continue only when visibility has returned to normal. It is better to bivouac for a night than to risk losing one's way by driving on blindly.

If you stray from the route, return to a beaconed location by following your tracks, not by going off in what you think is the direction of the lost route.

4 *Precautions in case of breakdown or getting lost*

If you break down or lose your way, above all REMAIN CALM, DO NOT PANIC and follow the essential rules given below:

a) NEVER ATTEMPT TO RETURN to a camp, or along a track which has been used, on FOOT even if you think that it is only a short distance. Remember that in general in these areas one always underestimates the distances and overestimates one's strength. It is also harder to spot a man on foot than a vehicle.
b) DO NOT MOVE AWAY FROM THE VEHICLE to look for shelter but protect yourself as much as possible from the weather either in or under the vehicle. In a hot desert area keep your clothes on in order to reduce dehydration. In a cold desert area wear your survival clothing.
c) RATION FOOD AND WATER from the time you lose your way and do not drink cooling liquid from the vehicles, since it may be toxic.
d) KEEP THE SIGNALLING DEVICES WITHIN REACH in order to be able to make your whereabouts known to any vehicle or plane as soon as you see or hear it; do not forget that it may not necessarily be able to see you. At night a fire or signals made by flashing headlights can be seen from a distance; alternatively, a lateral searchlight may be pointed, on narrow beam, towards the sky. If you have no rockets, mirrors and smoke can be spotted easily by day.

5 *Organization of searches*

It is essential, in order to have a chance of succeeding in a search, that the place of departure, probable date and time of arrival, destination and route are known for any vehicle or group of vehicles.

If one or more vehicles do not return, ie are 3 to 12h late depending on the nature and duration of the journey, the season and the weather conditions, a search will be carried out as follows:

PRELIMINARY SEARCH

Search the normal route following the beaconed track or road, using two vehicles, one with a radio, as far as the last known position of the missing vehicle(s), in the case of a vehicle which has broken down. If this is unsuccessful, the two vehicles should cover the route in the other direction, following tracks wherever possible or zigzagging periodically from side to side of the beaconed track. Do not go more than 3km away from the track without taking bearings and informing the base camp of the change of direction.

IN THE CASE OF AN UNSUCCESSFUL PRELIMINARY SEARCH

If the expedition has aircraft at its disposal (helicopters or planes), a search can be made by repeatedly flying over an area which extends beyond the normal route taken by the lost team.

If no aircraft are available, cover the same area with as many vehicles as possible, two vehicles covering each sector and maintaining close radio contact with the other vehicles and the base camp.

IN THE CASE OF AN UNSUCCESSFUL DETAILED SEARCH

If this search is unsuccessful ask for assistance from the civil and military authorities. Do not hesitate to cover the same routes twice because it is possible to pass very close to a vehicle without seeing it.

6 *Training of personnel*

These rules, based on the experience of numerous groups in desert zones and, unfortunately, on an analysis of a number of incidents and accidents of this kind, must be explained to all those participating in the expedition. At the start of the work the head of the expedition should discuss these instructions at a meeting attended by as large a number of staff as possible, stressing the points specific to the conditions of the particular expedition, and the importance of keeping to the safety rules. He may name an assistant leader responsible for seeing that the rules are kept.

During the course of the expedition the head, the group leaders and the prospectors will make sure that these instructions are strictly carried out, and may learn a great deal from their analysis of incidents which occur.

These instructions should be displayed in the cabs of all vehicles, together with a translation in the local language.

Appendix A3

Prospecting maps

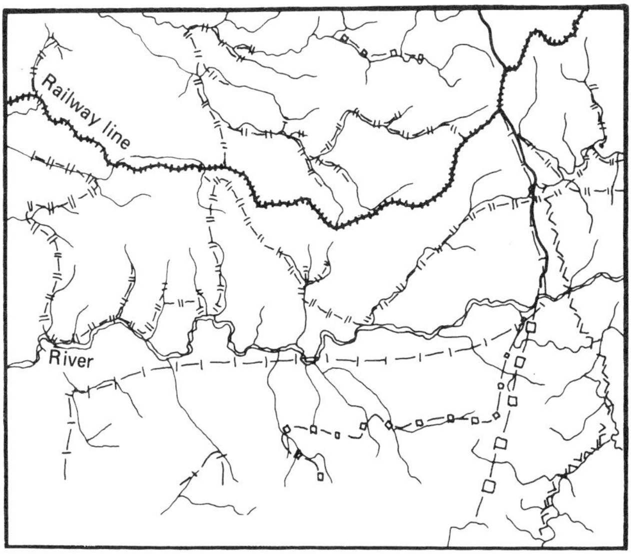

Map 1. *Map of the route, scale 1:200,000.*

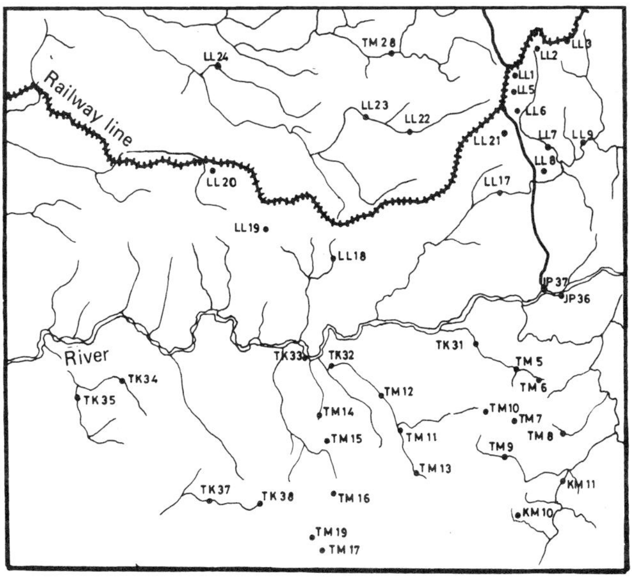

Map 2. *Map showing locations of petrographic samples, scale 1:200,000.*

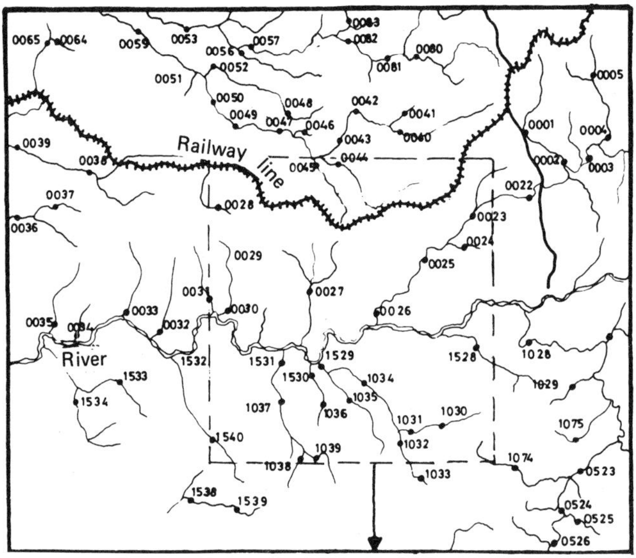

Map 3. *Map showing locations of alluvial samples, scale 1:200,000.*

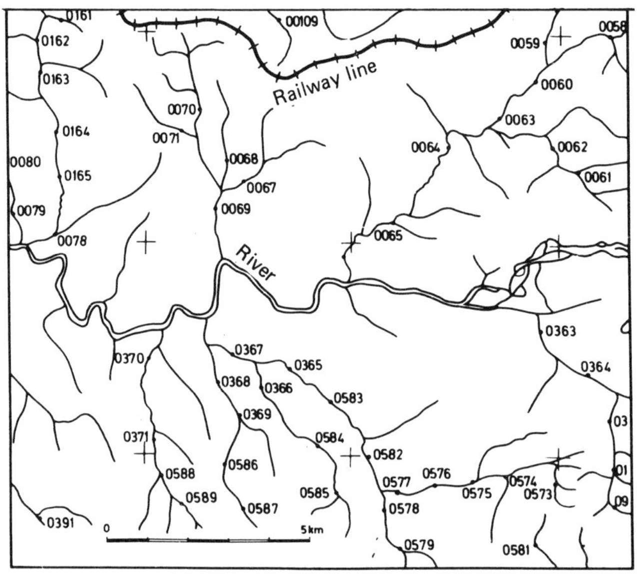

Map 4. *Map showing locations of geochemical samples.*

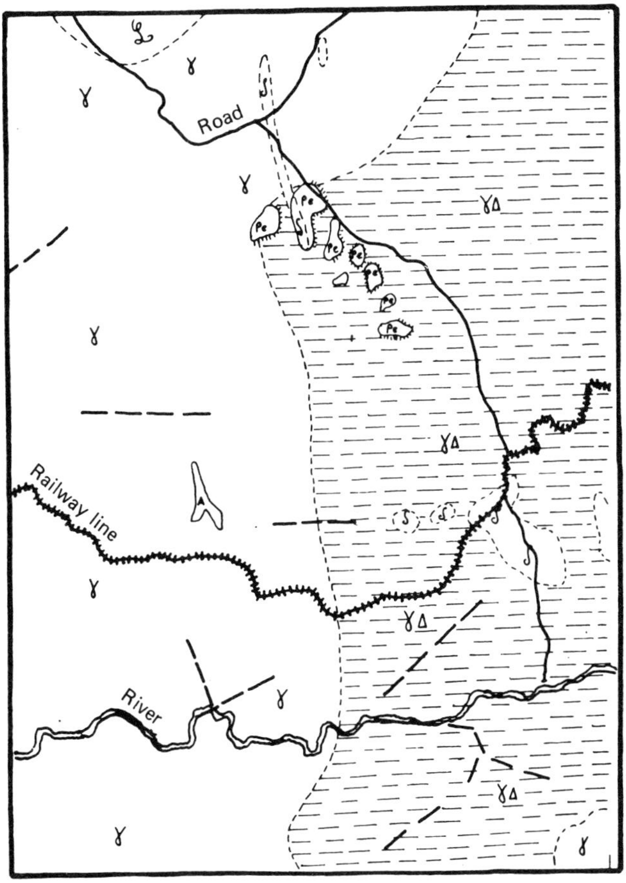

Map 5. *A geological reconnaissance map constructed from data from prospecting surveys (same sector, scale 1:200,000).*

Index